CMP

CRC

Flow Analysis Using a PC

H. Ninomiya and K. Onishi

Computational Mechanics Publications
Southampton Boston

Co-published with

CRC Press, Inc.
Boca Raton Ann Arbor Boston London

H. Ninomiya
Research Assistant, Master of Science
Department of Applied Physics
Fukuoka University
Fukuoka 814 - 01
Japan

K. Onishi
Department of Mathematics II
Science University of Tokyo
Wakamiya-cho 26
Shinjuku-ku
Tokyo 162
Japan

Co-published by Computational Mechanics Publications, Ashurst Lodge,
Ashurst, Southampton, SO4 2AA, UK and
Computational Mechanics, Inc., 25 Bridge Street, Billerica, MA 01821, USA

British Library Cataloguing-in-Publication Data
A Catalogue record for this title is available
from the British Library
ISBN 1-85312-144-4 Computational Mechanics Publications, Southampton, UK
ISBN 1-56252-077-6 Computational Mechanics Publications, Boston, USA
Library of Congress Catalog Card Number 91-72989

Published in USA and Canada by CRC Press Inc., 2000 Corporate Blvd.
N.W., Boca Raton, FL 33431

Library of Congress Cataloguing-in-Publication Data
Flow analysis using PC/ edited by Hiroshi Ninomiya and Kazuei Onishi
    p. cm.
    Translated from Japanese.
    Includes bibliographical references and index.
    ISBN 0-8493-7733-1
    1. Fluid mechanics–Data processing. 2. Numerical calculations.
3. Microcomputers. I. Ninomiya, Hiroshi, 1947- . II. Onishi,
Kazuei, 1947- .
QC151.F62 1991
532'.051'0285–dc20

91-24077
CIP

Contents

List of Figures   viii

List of Tables   xi

Base SI units used in the text   xii

Derived SI units from the base units   xii

PREFACE   xiii

CHAPTER 1. BASIC FLOW PROPERTIES   1
1.1 Fundamentals in Flow Analysis 1
     1.1.1 Streamlines and streamfunction 1
     1.1.2 Viscosity and vorticity 2
     1.1.3 Thermal effects 5
     1.1.4 Similarity in flows 6
1.2 Potential Flow 8
1.3 Viscous Fluid Flows 10
     1.3.1 Navier-Stokes equations 10
     1.3.2 Couette flow and Poiseuille flow 13
     1.3.3 Groundwater flow 14

CHAPTER 2. THE FINITE ELEMENT METHOD   19
2.1 Introduction to the Finite Element Method 19
     2.1.1 Solution process 19
     2.1.2 Ritz-Galerkin method 21
     2.1.3 Finite element subdivision 21
2.2 Interpolations 23
     2.2.1 Approximating functions 23
     2.2.2 One-dimensional interpolation 25
2.3 Method of Variations 27
     2.3.1 Ritz finite element method 29

2.4 Method of Weighted Residuals 32
   2.4.1 Galerkin finite element method 33
   2.4.2 Example in the Poiseuille flow 34
   2.4.3 Errors in one-dimensional approximations 38

CHAPTER 3. TWO-DIMENSIONAL PROBLEMS   47
3.1 Laplace Equation in Two Dimensions 47
3.2 Element Subdivision 48
   3.2.1 Triangulation 48
   3.2.2 Numbering of elements and nodes 48
   3.2.3 Topological data 51
3.3 Two-dimensional Linear Interpolation 51
3.4 Gauss-Green's Formula 55
3.5 Galerkin Finite Element Method 57
3.6 Total Equations 60
3.7 Errors in Two-dimensional Approximation 62

CHAPTER 4. POTENTIAL FLOWS   67
4.1 Basic Equations in Potential Flows 67
4.2 Discretisation of the Weak Form 68
4.3 Solution Procedure 70
4.4 Numerical Examples 72
   4.4.1 Flow around a circular cylinder 72
   4.4.2 Groundwater flow 75
4.5 Singularities 77

CHAPTER 5. TRANSIENT HEAT CONDUCTION   89
5.1 Basic Equations 89
5.2 Discretisation in Space and Time 90
5.3 Computational Procedure 93
5.4 Numerical Examples 98
   5.4.1 Heat conduction in a cooling cylinder 98
   5.4.2 Conduction in a composite material 100
5.5 Convergence of Approximate Solutions 102

CHAPTER 6. INCOMPRESSIBLE VISCOUS FLOW   109
6.1 Governing Equations 109
6.2 Finite Element Discretisation 111
6.3 Boundary Conditions of the Vorticity 114
6.4 Computational Scheme 115
6.5 Numerical Examples 120
   6.5.1 Flow facing a back step 120
   6.5.2 Viscous flow around a cylinder 122
   6.5.3 Cavity flow in a rectangle 125
   6.5.4 Rates of convergence 125

CHAPTER 7. THERMAL FLUID FLOW   **133**
7.1 Governing Equations 133
7.2 Finite Element Discretisation 135
7.3 Computational Scheme 138
7.4 Numerical Examples 140
    7.4.1 Natural convection in a closed compartment............... 140
    7.4.2 Bénard cell 143
    7.4.3 Forced thermal convection 144

CHAPTER 8. MASS TRANSPORT   **149**
8.1 Governing Equations 149
8.2 Finite Element Discretisation 151
8.3 Computational Scheme 153
8.4 Numerical Examples 155
    8.4.1 Smoke advection in the air 155
    8.4.2 Density-dependent viscous flow 156

CHAPTER 9. TIDAL CURRENT   **161**
9.1 Governing Equations 161
9.2 Finite Element Discretisation 163
9.3 Computational Scheme 165
9.4 Numerical Examples 167
    9.4.1 Travelling waves in a shallow channel 167
    9.4.2 Tidal current in the Ariake Bay 168

CHAPTER 10. PROGRAM INSTRUCTIONS   **175**
10.1 Program Specification 175
10.2 Data File Specification 177
10.3 Implementation Stream 179

Bibliography   **191**

Index   **193**

List of Figures

1.1   Movement of ink.   2
1.2   Forces acting in fluid at rest.   3
1.3   Laminar motion of viscous fluid.   4
1.4   Vortex motion in viscous fluid.   5
1.5   Buoyancy in thermal fluid.   6
1.6   Viscous flow at different Reynolds numbers.   7
1.7   Heat transport at different Rayleigh numbers.   9
1.8   Perfect fluid flow around a circular cylinder.   10
1.9   Tangential and normal components of velocity on boundary. . . .   13
1.10 One-dimensional viscous flows.   14

2.1   Finite element discretisation.   20
2.2   Finite element technique.   21
2.3   Approximate methods.   22
2.4   Element and node numbers.   23
2.5   Mesh and grid.   23
2.6   Linear interpolations.   24
2.7   Interpolation using roof functions.   26
2.8   Couette flow problem.   28
2.9   Finite element model in the Couette flow.   29
2.10 Finite element model in the Poiseuille flow.   35
2.11 Approximate solution of Poiseuille flow.   37

3.1   Some remarks on the element subdivision.   49
3.2   Node numberings.   50
3.3   Topology constituent in the finite elements.   52
3.4   A triangular element.   53
3.5   Linear triangular interpolation functions.   54
3.6   Integration paths.   55
3.7   Angles α, β.   57
3.8   A finite element neighbouring on the boundary.   59
3.9   A 4-element model.   60
3.10 Friedrichs-Keller mesh.   66

4.1   Flow diagrams.   71

4.2   Potential flow around a cylinder.   72
4.3   Finite element mesh.   74
4.4   Calculated flow around a circular cylinder using the velocity
      potential.   74
4.5   Calculated flow around a cylinder using the streamfunction. . . .   75
4.6   Seepage flow in a confined aquifer.   76
4.7   Mesh and boundary conditions for seepage flow.   76
4.8   Calculated seepage flow.   77
4.9   Pie-shaped domain with local polar coordinates.   78
4.10 Exterior domain to an airfoil.   82
4.11 Calculated flows around the NACA3409 airfoil.   84
4.12 Seepage through an earth dam.   85
4.13 Calculated equipotential lines.   85
4.14 Pie-shaped domain with mixed boundary conditions.   86
4.15 Perspective projection of the potential surface.   87

5.1   1/8 geometry of the cooling cylinder.   98
5.2   Finite element mesh.   99
5.3   Calculated transient heat conduction.   99
5.4   Quasi-steady and steady state solutions.   100
5.5   Core with the square cross section.   101
5.6   Evolution of calculated heat conduction.   101
5.7   Problem with internal singular point.   106
5.8   Uni-directional viscous fluid flow.   107
5.9   Calculated velocity contours.   107

6.1   Boundary conditions.   110
6.2   Triangular finite element.   112
6.3   Boundary condition for vorticity.   115
6.4   Computational scheme.   116
6.5   Geometry and boundary conditions.   121
6.6   Meshes.   121
6.7   Calculated streamlines in a back step channel.   122
6.8   Viscous flow around a cylinder.   123
6.9   Calculated viscous flow behind a circular cylinder;
      $Re = 100, t = 120s$.   124
6.10 Wind-driven cavity flow.   126
6.11 Calculated wind-driven cavity flows.   127
6.12 Horizontal velocity component on the vertical
      center line in steady state.   128
6.13 Parabolic flow over a half cylindrical obstacle.   130
6.14 Calculated flow over an obstacle.   131

7.1   Buoyancy induced by non-uniform temperature.   134
7.2   Cross section of an experimental vessel.   141
7.3   Calculated velocity vectors ($Ra = 1.89 \times 10^8$).   141

7.4 Calculated evolution of natural convection in the compartment ($Ra = 1.89 \times 10^8$). 142

7.5 Cross section of a shallow vessel. 143

7.6 Calculated Bénard convection in quasi-steady state ($Ra = 20250$). 143

7.7 Vertical channel between two parallel plates. 144

7.8 Calculated forced thermal convection near the heating spot ($t = 20\ s$). 145

7.9 Problem of natural convection. 146

7.10 Calculated results ($Pr = 1.0$). 146

7.11 Calculated profiles of streamfunction and temperature (black circles $Ra = 10^4$, white circles $Ra = 10^5$). 147

7.12 Problem of natural convection. 148

7.13 Calculated results ($Pr = 1.0$). 148

8.1 Buoyancy induced by non-uniform concentration. 150

8.2 Discharge of smoke into the atmosphere. 156

8.3 Finite element mesh. 156

8.4 Calculated convective diffusion of smoke ($Re = 828000$, $Pe = 0.918$). 157

8.5 Twin-cell vessel. 157

8.6 Calculated mixing in aquious solution ($Ra = 1.03 \times 10^{11}$). . . . 158

8.7 Coarse finite element mesh. 159

8.8 Calculated results ($t = 30\ s$) 159

9.1 Vertical cross section of the sea. 162

9.2 Long travelling waves in a channel. 167

9.3 Calculated elevation in the channel. 168

9.4 The Ariake Bay. 169

9.5 Calculated tides in the Ariake Bay. 170

9.6 Calculated current in the Ariake Bay. 171

9.7 Calculated tidal records in the Ariake Bay. 172

9.8 A small bay with an island. 173

9.9 Calculated tidal current ($t = 9\ hours$). 174

10.1 Potential flow problem. 186

10.2 Incompressible viscous flow. 187

10.3 Natural convection. 188

10.4 Air convective diffusion. 189

10.5 Tidal current. 190

List of Tables

1.1 Physical constants of water and air at 15°C, at 1 atmospheric
pressure.   4

1.2 Non-dimensional numbers.   8

2.1 Accuracy of the finite element solution.   37

3.1 Topological data.   52

6.1 Boundary values on the inlet. 131

Base SI units used in the text:

Quantity name	Unit name	Unit symbol
length	meter	m
mass	kilogram	kg
time	second	s
thermodynamic temperature	kelvin	K
(Supplementary unit)		
plane angle	radian	rad

Derived SI units from the base units:

Physical quantity	SI unit (symbol)	Definition for unit
force	newton (N)	$1N = 1kg \times 1m/s^2,\ kg\,m/s^2$
pressure	pascal (Pa)	$1Pa = 1N/m^2,\ kg\,/m\,s^2$
energy	joule (J)	$1J = 1Nm,\ kg\,m^2/s^2$
power	watt (W)	$1W = 1J/s,\ kg\,m^2/s^3$
(Customary units)		
pressure	standard atmosphere (atm)	$1atm = 101325 Pa$
Celsius temperature	degree Celsius ($°C$)	$\theta(°C) = T - 273.15(K)$
energy	calorie (cal)	$1cal\,(15°C) = 4.1855J$

PREFACE

This book is written expressly as a first course in numerical fluid mechanics. We have carefully selected fundamental and important flows of practical interest from various fluid flows and formulated them in the framework of the finite element method. Comparatively low cost personal computers such as the IBM-PC, NEC-PC, or more promising recent EWS (Engineering Work Stations) are assumed to be used. BASIC programs are provided on a diskette which accompanies the text. The aim of this book is to present a short introductory course on simple numerical modelling of fluid flows, using the elementary finite element method.

Flows of various degrees of complexity occur naturally. In this book, we shall mainly consider two-dimensional flows. We start our course with potential flows, which seem to be the simplest. The reader will be led easily through some typical examples in numerical computation of potential flow problems to the notion of flow velocity, velocity potential, streamfunction, and boundary conditions. The governing equation is the Laplace equation. As a starting point for transient analysis, we next consider unsteady heat conduction in solids. The corresponding governing equation is a parabolic equation. We then move to consideration of incompressible viscous fluid. The governing equations here are Navier-Stokes equations, which are expressed in terms of streamfunction and vorticity. The reader will learn related initial and boundary conditions. The discussion is extended to thermal convection due to buoyancy effects which include natural convection, forced convection, density-dependent convective diffusion, and the analysis of tidal currents in shallow sea water.

We confine ourselves to an elementary finite element method for the computational method. The minimum requisite of the finite element method is discussed in each chapter. We use the Ritz-Galerkin method as an introduction to the finite element method. Triangular 3-node linear elements are used exclusively for the two-dimensional analysis. These types of elements may not be the most efficient, but are the most familiar to beginners. This simple tool is valuable for solving the problems considered in this book. One should notice that some problems can be successfully formulated using only higher order finite elements in the flow analysis. Some special techniques will eventually lead to useful applications of the finite element method in computational fluid dynamics. However, detailed discussions are beyond the scope of this book. Interested readers are recommended to consult other more advanced references.

The best way to understand numerical modelling is to practise on available computers. Some sophisticated modelling concepts will reveal themselves through computation. Through practice, one can learn optimal mesh generation, appropriate boundary conditions, suitable time step size, *etc.* For this purpose, some numerical examples are included throughout the text as well as a series of exercises at the end of each chapter. As is often said, to write down mathematical expressions on a paper sheet is easy, but to obtain numerical solutions in practice is difficult. Experience of methods for numerical computation can be obtained through this work which should be very valuable.

A diskette of computer programs coded in BASIC is supplied with the text. They will run on standard PC (Personal Computers) and EWS. Perhaps it is worthwhile to note here that a 'computer program' is not synonymous with 'software'. 'Software' includes not only the programs but the technique of implementation. Unfortunately there are many people with much experience in programming but with no knowledge of how to input data, how to run the program, or even what was obtained from their output files. The users usually need more information than the programmer anticipated. To avoid such pitfalls, we offer in this book not only programs corresponding to respective flow problems, but also programs for automatic mesh generation, plotting of the mesh, renumbering of nodes as preprocessor, as well as to provide contour and vector plots of the numerical results as postprocessor.

The authors of this book are a physicist and an applied mathematician. Some useful programs they developed during their individual research activities were collected and edited in the form of a book. It was published in Japanese under the same title, and is the basis of this book. However, much new material and discussion has been added. All programs have been made compatible with IBM-PC. The revision is so extensive that we felt that we have virtually written a new book. Valuable advice and criticism from a wide variety of Japanese readers are cordially acknowledged and some of them were incorporated into this new edition.

Ninomiya and I wish to express our indebtedness to many people. We are most grateful to Mr Kenji Hayashi of Japan Weather Association for making the first Japanese edition of the book possible. We are equally grateful to Dr Carlos A. Brebbia of Computational Mechanics Institute in Southampton, England for his continuing encouragement. We also would like to express our appreciation to Dr Sebastian Koh for helpful suggestions and a careful reading of a part of the manuscript. Many thanks are due to Mrs Kinko Kobayashi, Mrs Yoko Ohura and my students for the preparation of the manuscript.

Kazuei Onishi

Chapter 1

BASIC FLOW PROPERTIES

To find numerical solutions of fluid flow problems, one requires basic equations governing the fluid flow. These equations are derived from laws of physics. In addition to the basic equations, one requires a knowledge of boundary conditions as well as physical properties of the fluid. In this chapter, we shall present some preliminaries for viscosity and thermal property of the fluid, which may strongly affect the flow behavior. We shall also present the basic flow equations and associated boundary conditions for two-dimensional flows in rectangular coordinate system.

1.1 Fundamentals in Flow Analysis

The three states of matter can be classified into a solid, liquid and gas. Liquid and gas are commonly called *fluids*. The main distinction between a liquid and a gas lies in their rate of change in the density. The density of gas changes more readily than that of liquid. However, they can be treated in the same way without taking into account the change of density, provided that the speed of flow is low as compared with the speed of sound propagating in the fluid. The fluid is called *incompressible* if the change of the density is negligible.

1.1.1 Streamlines and streamfunction

Suppose that ink is injected into a gently moving fluid. We can observe a streak of ink, as shown by the bold curve in Figure 1.1. The curve thus obtained is called a *streakline*. In general, the ink at the point B travelled there, not along the streakline but along a different curve, such as the one shown by the dotted line. This curve is called a *particle trajectory*.

We can consider yet another curve, which presents the flow pattern at the instant the ink reaches the point B. This curve is defined in such a way that the

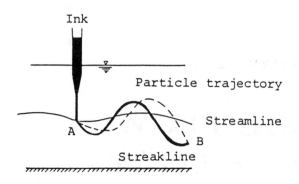

Figure 1.1: Movement of ink.

tangent to the curve and the flow velocity at the instant have the same direction at every point on the curve. The curve thus defined is called a *streamline*.

These curves coincide, only when the flow pattern does not change with time. In this case, the flow is called a *steady flow*. If the flow pattern changes with time, then it is called an *unsteady* or *transient flow*.

We denote by u and v the x- and y- components of the flow velocity (m/s), respectively. Since the flow vector with the components u, v at every point on the streamline has the same direction as the tangent vector at the same point, we have

$$\frac{dx}{u} = \frac{dy}{v} .$$

$$(1.1)$$

A planar curve can be generally expressed by means of a function in two variables. We consider the function $\psi(x, y)$ such that the relation: $\psi(x, y)$ = const., represents the streamlines. When ψ is a smooth function, the total derivative along the line must be equal to zero. Namely,

$$d\psi = \frac{\partial \psi}{\partial x} dx + \frac{\partial \psi}{\partial y} dy = 0 .$$

$$(1.2)$$

The function ψ is called a *streamfunction*. As a possible consequence, it follows from (1.1) and (1.2) that

$$u = \frac{\partial \psi}{\partial y} \quad , \quad v = -\frac{\partial \psi}{\partial x} .$$

$$(1.3)$$

These relations reveal an important fact that the partial derivatives of ψ form a vector which, after rotating 90° clockwise, coincides with the velocity vector.

1.1.2   Viscosity and vorticity

Let us consider forces acting in a fluid that is stationary in a vessel. The *gravity force* is acting in the isothermal fluid. Since the fluid is stationary, a counterforce must be acting in the fluid. This force is called *pressure* which will be denoted by

$p\,(Pa)$. Strictly speaking, the pressure is not a force. It is a force acting on any surface in the normal direction. The force per unit area (N/m^2) is called *stress*. Therefore, the pressure is a normal stress. Figure 1.2 illustrates the equilibrium of forces. Here we consider an imaginary cube in the fluid. Pressures acting on vertical surfaces balance out. The difference between the pressure acting on top and bottom surfaces exerts an upward force, known as *buoyancy force*, which balances with the gravity force. As the result, the fluid remains stationary.

The moving viscous fluid encounters an internal frictional force in the direction of motion due to the *viscosity*. This force is expressed as *shearing stress*, which acts on the unit area of a surface in the tangential direction. Consider a fluid at rest between two parallel plates. If one of the plates begins to move at constant velocity as shown in Figure 1.3, a motion is induced in the fluid. Through this process, the momentum is transported by the fluid due to its viscosity.

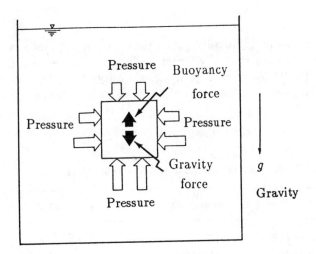

Figure 1.2: Forces acting in fluid at rest.

For most types of fluid, the shearing stress $\tau\,(Pa)$ is proportional to the negative of the spatial variation of the flow velocity du/dy, which is known as the *rate of strain*. Such fluid is called a *Newtonian fluid*. We have thus

$$\tau = -\mu \frac{du}{dy}, \tag{1.4}$$

where the proportionality constant μ is called the *viscosity coefficient* $(Pa \cdot s)$. The quotient of viscosity divided by the density $\rho\,(kg/m^3)$ of the fluid is called the *kinematic viscosity* (m^2/s), and it is written as

$$\nu = \mu/\rho. \tag{1.5}$$

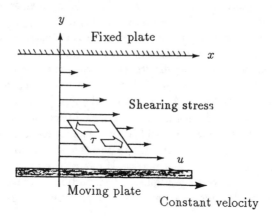

Figure 1.3: Laminar motion of viscous fluid.

The kinematic viscosity, rather than the viscosity, reflects the "sticky" motion of the fluid more faithfully. From Table 1.1 we can see that the viscosity of water is about 100 times as large as the viscosity of air, but the kinematic viscosity of water is smaller than that of air. This implies that air behaves more viscidly than water.

Table 1.1: Physical constants of water and air at 15°C, at 1 atmospheric pressure.

Material	constants	Water	Air
$\rho \ (kg/m^3)$	Density	999.1	1.225
$\mu \ (Pa \cdot s)$	Viscosity	1.14×10^{-3}	1.78×10^{-5}
$\nu \ (m^2/s)$	Kinematic viscosity	1.14×10^{-6}	1.45×10^{-5}
$\kappa \ (J/m \cdot s \cdot K)$	Heat conduction coefficient	5.9×10^{-1}	2.51×10^{-2}
$\lambda \ (m^2/s)$	Thermal conductivity	1.40×10^{-7}	2.02×10^{-5}
$\beta \ (1/K)$	Thermal expansion coefficient	1.5×10^{-4}	3.48×10^{-3}

Real fluids have various degrees of viscosity. The fluid in which the shearing stress is present is called a *viscous fluid*. If the shearing stress is negligibly small, then the fluid is considered *inviscid* and is called *perfect* or *ideal fluid*.

We consider the flow with spatial variations in the velocity as shown in Figure 1.4. If we immerse a small cogwheel into the fluid, it begins to rotate. At point A in the figure the cogwheel rotates counterclockwise, while at B, clockwise. We can characterize the rotational motion by introducing *vorticity*, defined as

$$\omega = \frac{\partial v}{\partial x} - \frac{\partial u}{\partial y} . \qquad (1.6)$$

The vorticity $\omega\,(1/s)$ is equal to twice the angular velocity of the cogwheel. We see that when $\omega > 0$ the cogwheel rotates counterclockwise, and when $\omega < 0$ it rotates clockwise. In particular when $\omega = 0$, we say that the flow is *irrotational*.

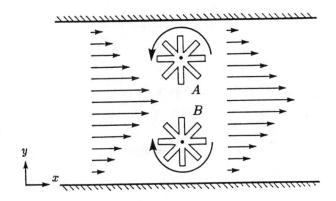

Figure 1.4: Vortex motion in viscous fluid.

1.1.3   Thermal effects

Thermal energy is transported in a fluid by conduction, convection and radiation. We shall only consider the conductive and convective heat transport in this book.

Suppose that the temperature is not uniform inside a fluid. Then the thermal energy is transported in the fluid so that the temperature would become uniform. If no mass transport is accompanied and only the heat is transported, then the phenomenon is called *heat conduction*. The rate of transport through unit area is called *flux*. The relationship between the *heat flux* $q\,(J/m^2{\cdot}s)$ and the temperature gradient is known as *Fourier's law*, which is described by

$$q = -\kappa\frac{\partial T}{\partial x} , \qquad (1.7)$$

where κ is called *heat conduction coefficient* $(J/m{\cdot}s{\cdot}K)$, and T is the temperature (K).

As the inhomogeneity of temperature distribution in the fluid increases, the balance between gravity force and buoyancy force is upset by the thermal expansion of the fluid. This is the starting point of *thermal convection*. In this case, the rate of heat transport is accelerated by the accompanying mass movement.

We illustrate this by considering a fluid at constant temperature T_0 in a vessel as shown in Figure 1.5. Suppose that the temperature of a portion of the fluid changes from T_0 to T. Then this portion of fluid receives the buoyancy force per unit mass (N/kg), given by

$$F = \beta g (T - T_0), \tag{1.8}$$

where g is the gravitational acceleration (m/s^2), and β is called the *thermal expansion coefficient* $(1/K)$. The buoyancy force F acts upward if $T > T_0$ and downward if $T < T_0$ with respect to the direction of the gravity.

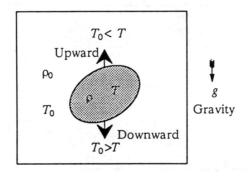

Figure 1.5: Buoyancy in thermal fluid.

1.1.4   Similarity in flows

By physical similarity we mean both geometric and kinematic similarities. Geometric similarity does not necessarily produce similarity in kinematics. Additional conditions are required for the physical similarity. These conditions are given generally by some non-dimensional numbers.

For the physical similarity in flows of viscous fluid, the corresponding non-dimensional number is known as *Reynolds number*, which is defined by

$$Re = \frac{U L}{\nu}, \tag{1.9}$$

where U is a characteristic velocity (m/s) in the flow system, L is the characteristic length (m), and ν is the kinematic viscosity. The flow fluctuates wildly as the Reynolds number Re becomes large, while the flow becomes gentle as Re becomes small. Figure 1.6 illustrates the evolution of viscous fluid flows at different Reynolds numbers. The flow is *laminar* at $Re \ll 1$. We can observe stable twin *vortices* just behind the plate for $Re \approx 10$. For $Re \approx 100$, these vortices grow and they are separated and carried downstream regularly by the main stream to form shedding known as *Kármán vortex shedding*. For larger Reynolds numbers, the flow becomes *turbulent*.

When thermal effects are present in the viscous fluid flow, we require two non-dimensional numbers to describe the similarity. These are known as the *Grashof number* and the *Prandtl number*, given respectively by

$$Gr \ = \ \frac{g\,\beta\,\Delta T\,L^3}{\nu^2}\,, \tag{1.10}$$

$$Pr \ = \ \nu/\lambda\,, \tag{1.11}$$

where ΔT is a characteristic temperature difference (K) in the flow system, and

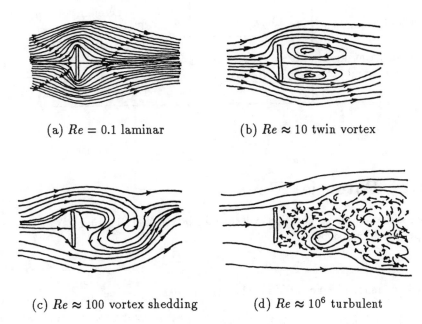

(a) $Re = 0.1$ laminar   (b) $Re \approx 10$ twin vortex

(c) $Re \approx 100$ vortex shedding   (d) $Re \approx 10^6$ turbulent

Figure 1.6: Viscous flow at different Reynolds numbers.

$\lambda = \kappa/\rho c$ being the *thermal conductivity* or *thermal diffusion coefficient* (m^2/s) and c is the *heat capacity* $(J/kg \cdot K)$.

Consider a heat conducting fluid at rest between two parallel plates, as illustrated in Figure 1.7. Suppose that the lower plate is heated uniformly. Then the mode of heat transport can be characterized by the *Rayleigh number*

$$Ra \ = \ Gr \cdot Pr \ = \ \frac{g\,\beta\,\Delta T\,L^3}{\lambda\,\nu}\,. \tag{1.12}$$

In the range of $Ra \leq 1700$, we can observe in the experiment that the fluid remains almost at rest. In this case, the heat is transported by conduction. In the range of $1700 \leq Ra \leq 50000$, a flow is induced. We can observe a series of regular cellular motions, known as *Bénard cells*. As the Rayleigh number becomes even larger, the regularity is lost and the flow becomes turbulent. Table 1.2 summarizes important non-dimensional numbers.

Table 1.2: Non-dimensional numbers.

Numbers	Definition	Physical constants	
Reynolds number	$Re = UL/\nu$	U	Velocity (m/s)
Péclet number	$Pe = UL/\lambda$	L	Length (m)
Grashof number	$Gr = g\beta\Delta T L^3/\nu^2$	ν	Kinematic viscosity (m^2/s)
Prandtl number	$Pr = \nu/\lambda$	g	Gravitational
			acceleration (m/s^2)
Schmidt number	$Sc = \nu/\eta$	β	Thermal expansion
			coefficient $(1/K)$
Rayleigh number	$Ra = Gr\,Pr$	ΔT	Temperature difference (K)
Froude number	$Fr = Re^2/Gr$	λ	Thermal conductivity (m^2/s)
		η	Diffusion coefficient (m^2/s)

1.2   Potential Flow

Some mathematical notations will now be introduced. The nabla ∇ is the vector operator, defined by

$$\nabla = (\frac{\partial}{\partial x}, \frac{\partial}{\partial y}),$$

and the inner product of ∇ is called the *Laplacian*. Namely,

$$\Delta = \nabla \cdot \nabla = \frac{\partial^2}{\partial x^2} + \frac{\partial^2}{\partial y^2}.$$

If we consider inviscid irrotaional flow, then from (1.6)

$$\omega = \frac{\partial v}{\partial x} - \frac{\partial u}{\partial y} = 0.$$

We know from vector calculus that the velocity vector with the components u, v can be derived from a scalar function as follows.

$$(u, v) = \nabla\Phi. \qquad (1.13)$$

The function Φ is called the *velocity potential* (m^2/s). If the fluid is incompressible, the *equation of continuity* reduces to

$$\frac{\partial u}{\partial x} + \frac{\partial v}{\partial y} = 0. \qquad (1.14)$$

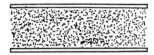

heated

(a) $Ra \leq 1700$

conductive

heated

(b) $1700 \leq Ra \leq 50000$

convective (Bénard cells)

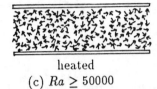

heated

(c) $Ra \geq 50000$

turbulent

Figure 1.7: Heat transport at different Rayleigh numbers.

After substitution of (1.13) into (1.14), we see that the velocity potential must satisfy the *Laplace equation*

$$\nabla^2 \Phi = 0 . \qquad (1.15)$$

Since the solution of the Laplace equation can be expressed mathematically by classical potentials, the flow governed by the Laplace equation is called *potential flow*.

The potential flow can be formulated alternatively in terms of the streamfunction ψ. When we substitute (1.3) into (1.6), the irrotational condition gives us another Laplace equation:

$$\nabla^2 \psi = 0 . \qquad (1.16)$$

As an example of the potential flow, we consider an ideal fluid flow around a circular cylinder. Let an infinitely long cylinder with radius R be submerged in the fluid, which is flowing at the uniform velocity $(u, v) = (U, 0)$ as illustrated in Figure 1.8. Using the polar coordinate system (r, θ) with the origin taken at the axis of the cylinder, we can see that the corresponding velocity potential and streamfunction are given respectively as follows.

$$\Phi = U \left(r + \frac{R^2}{r} \right) \cos \theta , \qquad (1.17)$$

$$\psi = U \left(r - \frac{R^2}{r} \right) \sin \theta . \qquad (1.18)$$

In Figure 1.8, streamlines are depicted by solid curves. *Equi-potential lines* are depicted by dotted lines. These lines intersect at right angles, and constitute an orthogonal net.

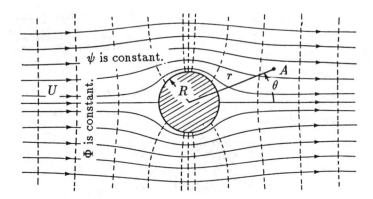

Solid curves: streamline, dotted curves: equi-potential lines.

Figure 1.8: Perfect fluid flow around a circular cylinder.

1.3   Viscous Fluid Flows

We shall consider basic equations governing flow of a viscous fluid.

1.3.1   Navier-Stokes equations

Shearing stresses are present due to the viscosity inherent in the real fluid. For the Newtonian fluid, equation (1.4) can be extended to give the following relationships between stresses and rate of strains.

$$
\begin{pmatrix} \tau_{xx} & \tau_{xy} \\ \tau_{yx} & \tau_{yy} \end{pmatrix} = \begin{pmatrix} p - 2\mu\frac{\partial u}{\partial x} & -\mu(\frac{\partial v}{\partial x} + \frac{\partial u}{\partial y}) \\ -\mu(\frac{\partial u}{\partial y} + \frac{\partial v}{\partial x}) & p - 2\mu\frac{\partial v}{\partial y} \end{pmatrix} , \tag{1.19}
$$

where $\tau_{xx}, \tau_{xy}, \tau_{yx}, \tau_{yy}$ are components of the stress tensor.

To derive the formulas expressing the acceleration of the fluid motion, we consider a fluid particle at position (x, y), which is moving with the velocity $u(x, y, t)$ and $v(x, y, t)$, at the time t. Suppose that the fluid particle moves to position $(x + u\Delta t, y + v\Delta t)$ in the time interval Δt. Then the change in velocity is given by the following expressions.

$$
\begin{aligned}
\Delta u &= u(x + u\Delta t, y + v\Delta t, t + \Delta t) - u(x, y, t) \\
&= \frac{\partial u}{\partial x} u\Delta t + \frac{\partial u}{\partial y} v\Delta t + \frac{\partial u}{\partial t} \Delta t + O(\Delta t^2) . \\
\Delta v &= v(x + u\Delta t, y + v\Delta t, t + \Delta t) - v(x, y, t) \\
&= \frac{\partial v}{\partial x} u\Delta t + \frac{\partial v}{\partial y} v\Delta t + \frac{\partial v}{\partial t} \Delta t + O(\Delta t^2) .
\end{aligned}
$$

The rate of change of the velocity in the flow direction, which we denote by the

Lagrange derivatives Du/Dt and Dv/Dt, is given respectively by

$$\frac{Du}{Dt} = \lim_{\Delta t \to 0} \frac{\Delta u}{\Delta t} = \frac{\partial u}{\partial t} + u\frac{\partial u}{\partial x} + v\frac{\partial u}{\partial y} ,$$

$$\frac{Dv}{Dt} = \lim_{\Delta t \to 0} \frac{\Delta v}{\Delta t} = \frac{\partial v}{\partial t} + u\frac{\partial v}{\partial x} + v\frac{\partial v}{\partial y} .$$

The equation of continuity and equations of motion are basic equations for the viscous fluid flow. The equation of continuity is derived from the law of conservation of mass. The equations of motion are derived from the law of conservation of momentum. Under the assumption of incompressibility, we have *the equation of continuity* in the form

$$\frac{\partial u}{\partial x} + \frac{\partial v}{\partial y} = 0 , \qquad (1.20)$$

and the *equations of motion*

$$\rho\frac{Du}{Dt} + \frac{\partial \tau_{xx}}{\partial x} + \frac{\partial \tau_{xy}}{\partial y} = \rho F_x , \qquad (1.21)$$

$$\rho\frac{Dv}{Dt} + \frac{\partial \tau_{yx}}{\partial x} + \frac{\partial \tau_{yy}}{\partial y} = \rho F_y , \qquad (1.22)$$

where F_x, F_y are external forces per unit mass $(N/kg = m/s^2)$. The leading term on the left-hand side of each equation of motion denotes the product of mass per unit volume and acceleration, *i.e.* momentum per unit volume. The rest of the terms denote forces exerted on the fluid per unit volume. This implies that the equations of motion have the familiar form $m\alpha = f$ in Newton's second law of motion.

Substituting (1.19) into the equations of motion, and utilizing (1.20), one can obtain the following *Navier-Stokes equations*.

$$\frac{\partial u}{\partial t} + u\frac{\partial u}{\partial x} + v\frac{\partial u}{\partial y} = -\frac{1}{\rho}\frac{\partial p}{\partial x} + \nu \nabla^2 u + F_x , \qquad (1.23)$$

$$\frac{\partial v}{\partial t} + u\frac{\partial v}{\partial x} + v\frac{\partial v}{\partial y} = -\frac{1}{\rho}\frac{\partial p}{\partial y} + \nu \nabla^2 v + F_y . \qquad (1.24)$$

The first term on the left-hand side of each equation denotes a component of the acceleration at a point in the flow field. If they both vanish, the flow is in a *steady state*. The sum of second and third terms on the left-hand side of each equation is called *convective term*. The first terms on the right-hand side of these equations denote the *pressure gradient*, which serves as a driving force of the fluid motion. The second terms on the right-hand side are *viscosity terms*. When the magnitude of the viscosity terms are large as compared to the convective terms, the velocity of the flow varies gently. When the magnitude of the viscosity term is small however, the flow tends to rustle and has large variations in its velocity.

Suppose for the moment that $F_x = 0$ and $F_y = 0$. One can eliminate the pressure terms in the two-dimensional Navier-Stokes equations. To this end, one differentiates (1.24) with respect to x and (1.23) with respect to y, and subtracts one from another. Using the definition of the vorticity (1.6) and the equation of continuity (1.20), one obtains the following *equation of vorticity transport*.

$$\frac{\partial \omega}{\partial t} + u\frac{\partial \omega}{\partial x} + v\frac{\partial \omega}{\partial y} = \nu \nabla^2 \omega . \qquad (1.25)$$

Moreover, substituting (1.3) into (1.6), one obtains

$$\nabla^2 \psi = -\omega , \qquad (1.26)$$

which relates the vorticity to the steamfunction. The velocity components can be calculated from the streamfunction via (1.3).

Generally speaking, there are two ways to analyze the viscous fluid flow numerically. One is the direct way, in which the velocity components u, v and the pressure p are employed as unknowns in the numerical analysis. This is called the *primitive variable approach*. Another is the indirect way, in which streamfunction ψ and vorticity ω are employed as unknowns. In this book we shall consider the *streamfunction-vorticity approach*, which often suits the numerical computation of two-dimensional viscous fluid flow.

Boundary conditions are required to determine the solution of equations uniquely. Typical boundary conditions for equation (1.26) are;

$$\psi = \psi_B \quad \text{on} \quad \Gamma_\psi , \qquad (1.27)$$

$$\frac{\partial \psi}{\partial n} = -V_s \quad \text{on} \quad \Gamma_s , \qquad (1.28)$$

where Γ_ψ is the part of the boundary of the flow domain Ω, along which the value of the streamfunction ψ_B is prescribed, and Γ_s is the rest of the boundary, along which the value of the tangential velocity V_s is prescribed, as shown in Figure 1.9. We notice here that Ω depicted in the figure is a multiply connected domain. The boundary components are always oriented so that the domain is on the left-hand side of the boundary.

Typical boundary conditions for equation (1.25) are;

$$\omega = \omega_B \quad \text{on} \quad \Gamma_\omega , \qquad (1.29)$$

$$\frac{\partial \omega}{\partial n} = \chi_B \quad \text{on} \quad \Gamma_\chi , \qquad (1.30)$$

where Γ_ω is the part of the boundary, along which the value of the vorticity ω_B is prescribed, and Γ_χ is the rest of the boundary, along which the value of the normal derivative χ_B is prescribed. Other important types of boundary conditions on the vorticity will be presented in section 6.3.

The boundary condition of the type (1.27) or (1.29), where the boundary values of unknowns are specified, is called a *Dirichlet condition*. The boundary condition of the type (1.28) or (1.30), where the values of normal derivatives are specified, is called a *Neumann condition*.

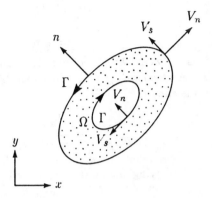

Figure 1.9: Tangential and normal components of velocity on boundary.

1.3.2 Couette flow and Poiseuille flow

We consider a flow whose Reynolds number is so small that, in the Navier-Stokes equations, the convective terms are negligible as compared to the viscosity terms. Then we have the following *Stokes equations*.

$$\frac{\partial u}{\partial t} \;=\; -\frac{1}{\rho}\frac{\partial p}{\partial x} + \nu\,\nabla^2 u + F_x\,, \qquad (1.31)$$

$$\frac{\partial v}{\partial t} \;=\; -\frac{1}{\rho}\frac{\partial p}{\partial y} + \nu\,\nabla^2 v + F_y\,. \qquad (1.32)$$

Consider a one-dimensional flow between a wall and a parallel plate at a distance L from it, as shown in Figure 1.10(a). The wall A in the figure is fixed, while the plate B is moving, with constant velocity U, in a channel which is parallel to the wall. The flow induced in the channel is called *Couette flow*. In this case, pressure gradient in the x direction is not induced. Namely,

$$\frac{\partial p}{\partial x} \;=\; 0\,.$$

In the steady state, the corresponding Stokes equation (1.31) is simplified to

$$0 \;=\; \nu\,\frac{d^2 u}{dy^2}\,. \qquad (1.33)$$

The general solution of this differential equation is given by

$$u \;=\; C_1\,y + C_2 \qquad (1.34)$$

with two arbitrary constants C_1 and C_2. To determine the constants, we consider the following boundary conditions.

$$u \;=\; 0 \quad \text{at} \quad y = 0\,, \quad \text{and} \quad u = U \quad \text{at} \quad y = L\,.$$

The particular solution satisfying these boundary conditions is given by

$$u = U \frac{y}{L}.$$

(1.35)

We now consider a flow induced by the given pressure gradient $\alpha = -dp/dx$ only in the x direction between two fixed parallel walls, as shown in Figure 1.10(b). The flow is called *Poiseuille flow*. In the steady state, the corresponding Stokes equation (1.31) is simplified to

$$0 = \frac{\alpha}{\rho} + \nu \frac{d^2 u}{dy^2}.$$

The particular solution satisfying the boundary conditions $u = 0$ at both $y = 0$ and $y = L$ is given by

$$u = \frac{\alpha}{2\mu} y (L - y).$$

(1.36)

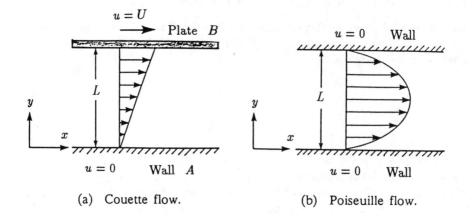

(a) Couette flow.   (b) Poiseuille flow.

Figure 1.10: One-dimensional viscous flows.

1.3.3   Groundwater flow

Rain and melted snow infiltrate into the ground. They penetrate through pores of the soil and become the groundwater. The soil stratum saturated with the groundwater is called *aquifer*. The groundwater moves through the aquifer as *seepage flow*. Velocity of the groundwater motion is defined as the rate of volume (m^3/s) divided by the cross section (m^2) consisting of the soil and pores of the soil matrix. The fluid is assumed to be incompressible. Stokes equations apply, because the velocity of the seepage flow is often small.

With y axis directed upward in the opposite direction of the gravitational acceleration, the Stokes equations can be written as follows.

$$\frac{\partial u}{\partial t} = -\frac{1}{\rho}\frac{\partial p}{\partial x} + \nu\nabla^2 u$$

$$= -g\frac{\partial}{\partial x}\left(\frac{p}{g\rho}\right) + \nu\nabla^2 u \,, \tag{1.37}$$

$$\frac{\partial v}{\partial t} = -\frac{1}{\rho}\frac{\partial p}{\partial y} + \nu\nabla^2 v - g$$

$$= -g\frac{\partial}{\partial y}\left(\frac{p}{g\rho} + y\right) + \nu\nabla^2 v \,. \tag{1.38}$$

Let $H = p/(g\rho) + y$. This is known as the *piezometric head* (m). It is the sum of *pressure head* and *elevation* from the referenced horizontal plane.

We assume that the aquifer is hydraulically isotropic. The viscosity terms are often proportional to the seepage velocity. Namely,

$$\nu\nabla^2 u = -\frac{g}{k}u \,, \tag{1.39}$$

$$\nu\nabla^2 v = -\frac{g}{k}v \,, \tag{1.40}$$

where k is called *hydraulic conductivity* (m/s). Substituting (1.39) and (1.40) into (1.37) and (1.38) respectively, we have

$$\frac{\partial u}{\partial t} = -g\left(\frac{\partial H}{\partial x} + \frac{u}{k}\right) \,, \tag{1.41}$$

$$\frac{\partial v}{\partial t} = -g\left(\frac{\partial H}{\partial y} + \frac{v}{k}\right) \,. \tag{1.42}$$

In the steady state, we arrive at the following *D'Arcy law*:

$$u = -k\frac{\partial H}{\partial x} \quad , \quad v = -k\frac{\partial H}{\partial y} \,. \tag{1.43}$$

We now introduce hydraulic potential $\Phi = -kH$. The D'Arcy law becomes

$$u = \frac{\partial \Phi}{\partial x} \quad , \quad v = \frac{\partial \Phi}{\partial y} \,. \tag{1.44}$$

Combining these with the equation of continuity (1.14), we obtain the Laplace equation

$$\nabla^2\Phi = 0 \,.$$

Thus we can treat the ground water flow as a potential flow.

Exercises

1.1 As a consequence of (1.1) and (1.2), we can obtain another possibility:

$$u = -\frac{\partial \psi}{\partial y} \quad , \quad v = \frac{\partial \psi}{\partial x}$$

rather than (1.3). Explain the change in properties of the streamfunction, if we should employ this definition.

1.2 Under the assumption of incompressibility, derive the equation of continuity (1.14) from the general continuity equation

$$\frac{\partial \rho}{\partial t} + \frac{\partial}{\partial x_j}(\rho v_j) = 0 .$$

Here, we use the *Einstein's summation convention* for repeated index j.

1.3 The Laplace equation (1.15) in polar coordinate system with $x = r \cos \theta$, $y = r \sin \theta$ is written as

$$\frac{1}{r}\frac{\partial}{\partial r}(r\frac{\partial \Phi}{\partial r}) + \frac{1}{r^2}\frac{\partial^2 \Phi}{\partial \theta^2} = 0 .$$

Derive the solution (1.17) subject to the boundary conditions

$$(\frac{\partial \Phi}{\partial r})_{r=R} = 0 \quad , \quad (\frac{\partial \Phi}{\partial r})_{r \to \infty} = U \cos \theta$$

by the method of separation of variables $\Phi(r, \theta) = P(r)\Theta(\theta)$.

Hint : From the boundary conditions, we have

$$(\frac{dP}{dr})_{r=R} = 0 \quad , \quad (\frac{dP}{dr})_{r \to \infty} = 1 \quad , \quad \Theta = U \cos \theta .$$

From the Laplace equation, we have

$$\frac{1}{r}\frac{d}{dr}(r\frac{dP}{dr})\Theta + \frac{1}{r^2}P\frac{d^2\Theta}{d\theta^2} = 0 .$$

By rearranging the terms, we can obtain

$$\frac{r\frac{d}{dr}(r\frac{dP}{dr})}{P} = -\frac{\frac{d^2\Theta}{d\theta^2}}{\Theta} = \lambda .$$

Since the most left-hand side is a function of r only, and since the second is a function of θ only, we see that λ must be a constant. It follows that

$$r\frac{d}{dr}(r\frac{dP}{dr}) - \lambda P = 0 \quad , \quad \frac{d^2\Theta}{d\theta^2} + \lambda\Theta = 0 .$$

To determine the value of λ, notice that Θ must be a periodic function with the period 2π: $\Theta(0) = \Theta(2n\pi), n = \pm 1, \pm 2, \ldots$. By integration by parts, we see

$$\int_0^{2n\pi} (\frac{d^2\Theta}{d\theta^2} + \lambda\Theta)\,\Theta\,d\theta$$

$$= -\int_0^{2n\pi} (\frac{d\Theta}{d\theta})^2\,d\theta + \lambda \int_0^{2n\pi} \Theta^2\,d\theta = 0 \ .$$

This implies that $\lambda > 0$. Put $\lambda = m^2$. The general solution for Θ is given by

$$\Theta = A_m \cos m\theta + B_m \sin m\theta \ .$$

A particular solution satisfying the boundary condition is given by

$$\Theta = U \cos\theta \quad \text{with} \quad m = 1 \ .$$

Then the equation for P becomes

$$r\frac{d}{dr}(r\frac{dP}{dr}) - P = 0 \ .$$

Put $P = Cr^\rho$ with arbitrary constant C. We can see that $\rho = \pm 1$. The general solution for P is given by

$$P = (C_1 r + \frac{C_2}{r}) \ .$$

From the boundary conditions for $dP/dr = C_1 - C_2/r^2$, it follows that $C_1 = 1$ and $C_2 = R^2$.

1.4 Derive the solution (1.36) directly from the equation of continuity (1.20) and the Navier-Stokes equations (1.23), (1.24).

Hint : The assumptions of the problem read

$$\frac{\partial u}{\partial t} = 0 \ , \quad \frac{\partial v}{\partial t} = 0 \quad \text{(in steady state)} ,$$

$$v = 0 \quad \text{(one-dimensional flow)} ,$$

$$-\frac{\partial p}{\partial x} = \alpha = const. \quad \text{(given pressure gradient)} .$$

From the equation of continuity, we have

$$\frac{\partial u}{\partial x} = 0 \ .$$

This implies that $u = u(y)$, a function of y only. From the Navier-Stokes equations without external forces, we have

$$u\frac{\partial u}{\partial x} = -\frac{1}{\rho}\frac{\partial p}{\partial x} + \nu \nabla^2 u \quad , \quad 0 = -\frac{1}{\rho}\frac{\partial p}{\partial y} \ .$$

It follows that

$$0 = \frac{\alpha}{\rho} + \nu \frac{d^2 u}{dy^2} \quad \text{and} \quad p = p(x) .$$

The boundary condition $u = 0$ at both $y = 0$ and $y = L$ yields (1.36).

1.5 Using any characteristic length L and velocity U of the flow problem under consideration, show that following new variables with asterisk * are non-dimensional variables.

$$x^* = \frac{x}{L} \quad , \quad y^* = \frac{y}{L} \quad , \quad t^* = \frac{U}{L} t ,$$

$$u^* = \frac{u}{U} \quad , \quad v^* = \frac{v}{U} \quad , \quad p^* = \frac{p}{U^2 \rho} .$$

In terms of these variables, show that the equation of continuity (1.20) and the Navier-Stokes equations (1.23), (1.24) without external forces can be expressed in the following non-dimensional forms.

$$\frac{\partial u^*}{\partial x^*} + \frac{\partial v^*}{\partial y^*} = 0 ,$$

$$\frac{\partial u^*}{\partial t^*} + u^* \frac{\partial u^*}{\partial x^*} + v^* \frac{\partial u^*}{\partial y^*} = -\frac{\partial p^*}{\partial x^*} + \frac{1}{Re} \left(\frac{\partial^2 u^*}{\partial x^{*2}} + \frac{\partial^2 u^*}{\partial y^{*2}} \right) ,$$

$$\frac{\partial v^*}{\partial t^*} + u^* \frac{\partial v^*}{\partial x^*} + v^* \frac{\partial v^*}{\partial y^*} = -\frac{\partial p^*}{\partial y^*} + \frac{1}{Re} \left(\frac{\partial^2 v^*}{\partial x^{*2}} + \frac{\partial^2 v^*}{\partial y^{*2}} \right) .$$

This shows that only the Reynolds number is involved, which characterizes the similarity of Newtonian fluid flows.

1.6 Show that the groundwater flow in unsteady state is expressed by the parabolic equation

$$S \frac{\partial \Phi}{\partial t} = k \nabla^2 \Phi$$

with the *storage coefficient* $S \, (1/m)$.

Chapter 2

THE FINITE ELEMENT METHOD

In this chapter we shall present the basic knowledge required for the implementation of the numerical solution of boundary value problems by finite element methods. The Ritz-Galerkin method using finite elements is applied to both Couette and Poiseuille flows in one dimension in order to show in detail the approximation process involved in the method. The mathematical aspect is briefly discussed.

2.1 Introduction to the Finite Element Method

We shall take a glance at the finite element method in this section. The numerical solution process associated with the method is simple. The method which we shall consider is equivalent to the Ritz-Galerkin method with finite elements as the trial functions.

2.1.1 Solution process.

The finite element method is a numerical procedure for approximate solutions of boundary value problems. Continuum, which corresponds to a domain of analysis, is replaced by a finite series of small segments with a finite number of degrees of freedom in such a way that local behaviour of the continuum is confined to each segment.

For one- and two-dimensional boundary value problems, the approximate solution proceeds as shown in Figure 2.1. For example, the domain of analysis is subdivided into a finite number of small linear segments or triangles. A finite number of points referring to these segments are called *nodes*.

The exact solution can be expressed approximately in terms of the nodal values and interpolation functions within the segments. Piecewise polynomials of relatively low orders or splines have been used as the interpolation functions.

The small segments with these associated nodes and interpolation functions are called the *finite elements*.

A small algebraic system of equations with unknown nodal values is derived from each finite element by the Ritz-Galerkin method. If the boundary value problem is linear, then the corresponding algebraic system is also linear. This small system of equations is called the *element equation*. Matrices associated with the element equations are called the *element matrices*. In the finite element method, it is important to notice that the discretisation is carried out elementwise.

All the element equations are collected together to form the *global* or *total equations*. The collection process is called *assembly*. The matrices obtained after the assembly are called *assembled matrices*.

A large system of the global equations is solved numerically. If the system is linear, the Gaussian elimination method is often used. The solution vector contains approximate nodal values to the exact solution as its components.

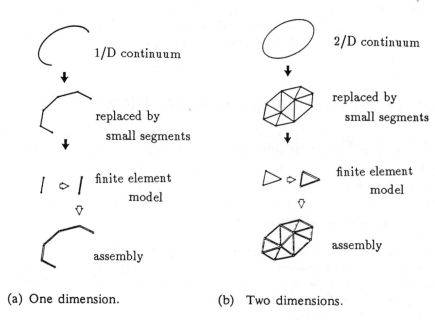

1/D continuum   2/D continuum

replaced by small segments

finite element model

assembly

(a) One dimension.   (b) Two dimensions.

Figure 2.1: Finite element discretisation.

One of the outstanding characteristics of the finite element method lies in its consistency in the course from the mathematical formulation to the computer programming. The software configurations of the method are almost common irrespective of the problems to be solved. The finite element programs can be coded in highly modular forms, and they are thus easy to read and write as shown in Figure 2.2.

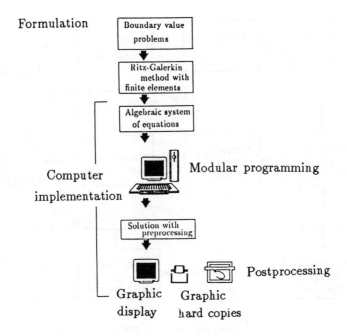

Figure 2.2: Finite element technique.

2.1.2 Ritz-Galerkin method.

The *Ritz method* (or *Rayleigh-Ritz method*) is an approximate technique for finding a minimum of functionals, based on the method of variations. The *Galerkin method*, on the other hand, is also an approximate technique that is used to find solutions to boundary value problems. It is based on the method of weighted residuals. To make the numerical implementation easy, piecewise polynomials are used for the interpolation functions and weighting functions in both the Ritz and the Galerkin methods. Since the interpolation function does not vanish only on a few of the finite elements, the method has come to be called the *Ritz-Galerkin finite element method.*

When the physical problem under consideration has a functional to be minimized, the Ritz and Galerkin methods are equivalent. However, when the physical problem has no such functionals, the Ritz method is not applicable. The Galerkin method can be applied even for such problem, as shown in Figure 2.3.

2.1.3 Finite element subdivision.

We consider a two-dimensional domain of analysis, subdivided into a finite number of small triangles. Three vertices of one of the triangles are taken as referring nodes. Smooth functions defined on the triangle can be interpolated approxi-

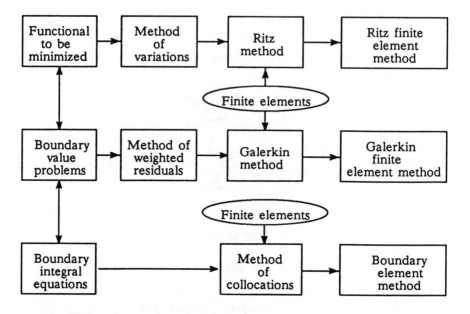

Figure 2.3: Approximate methods.

mately by using those nodal values. Triangles related to a set of interpolation functions are called the *triangular finite elements*. However, the triangles alone have been customarily called finite elements.

The finite elements and nodes are numbered consecutively. The numbers assigned to the elements are called *element numbers*. The numbers assigned to the nodes are called *node numbers*, see Figure 2.4. The three element numbers attached to a triangular element are read counterclockwise for convenience sake.

The collection of the finite elements is called a *mesh*. It is recommended that the region, where the solution is expected to behave with large variations, is subdivided into a *fine mesh*. It is also recommended that the region, where the solution is expected to behave smoothly with slight variations, is subdivided into a *coarse mesh*. These tricks not only reduce the total number of elements, but also are expected to improve the accuracy. In an ironic sense, it had been often said that the finite element subdivision cannot be done before the solution has been obtained! After the interpolation function have been chosen, one should keep in mind that the ultimate quality of the finite element solution solely depends on the mesh. In this sense, the mesh distribution is most important. Recently, the situation has been improved by the technique of adaptive remeshing.

The domain with curved boundaries can be approximated well to some extent by the triangular mesh, as illustrated in Figure 2.5. Relatively smooth patterns of the flow velocity can be obtained near the boundary using the curve fitting grid as compared to ones using the rectangular grid.

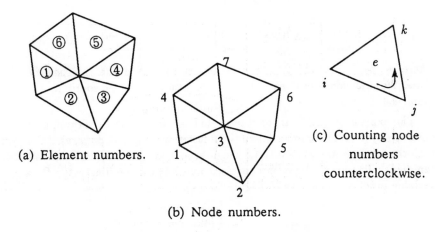

(a) Element numbers.

(b) Node numbers.

(c) Counting node
numbers
counterclockwise.

Figure 2.4: Element and node numbers.

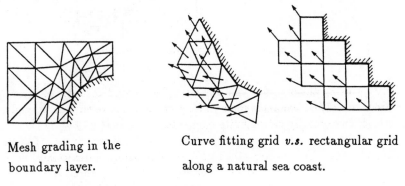

Mesh grading in the
boundary layer.

Curve fitting grid *v.s.* rectangular grid
along a natural sea coast.

Figure 2.5: Mesh and grid.

2.2   Interpolations

Functions that vary smoothly can be represented in terms of their nodal values
by using suitably chosen interpolation functions. As an introduction to the
theory of interpolations, we shall confine ourselves to the Lagrange interpolation
in one and two dimensions.

2.2.1   Approximating functions.

Before we consider the approximation of a single-variable function in detail,
we shall consider a two-variable function $u(x, y)$ defined on the domain Ω, as
shown in Figure 2.6 (b). Suppose that the domain is subdivided using n nodes
into a series of triangular finite elements. The function is approximated by a
polyhedral surface resembling a mosaic. The approximate function $\hat{u}(x, y)$ can

be constructed in terms of the nodal values $u_1, u_2, \ldots, u_n$ as follows.

$$\hat{u}(x, y) = u_1\phi_1(x, y) + u_2\phi_2(x, y) + \cdots + u_n\phi_n(x, y), \qquad (2.1)$$

where $\phi_1, \phi_2, \ldots, \phi_n$ are two-variable functions, called the *interpolation functions*. We shall consider linear interpolation functions in the coordinate variables x and y.

The expression (2.1) can be interpreted to mean that the function $\hat{u}(x, y)$ is obtained by a linear combination of the n functions $\phi_1(x, y), \phi_2(x, y), \ldots, \phi_n(x, y)$ with the coefficients $u_1, u_2, \ldots, u_n$, the interpolation functions are also called *base functions*. The approximation expressed by (2.1) can be illustrated for a single variable function, as shown in Figure 2.6 (a).

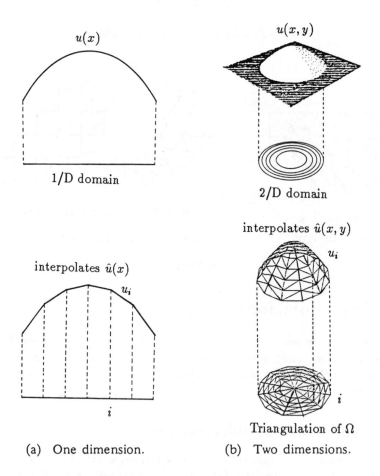

(a) One dimension.   (b) Two dimensions.

Figure 2.6: Linear interpolations.

2.2.2   One-dimensional interpolation.

We shall consider a single-variable function $u(x)$ in the interval $0 \le x \le 1$. Let the interval be divided into four subintervals with equal length, as shown in Figure 2.7 (a). Let h be the common length of each subinterval. We put

$$
\begin{array}{lllll}
u_1 & = & u(x_1) & \text{at} & x_1 = 0 , \\
u_2 & = & u(x_2) & \text{at} & x_2 = h , \\
u_3 & = & u(x_3) & \text{at} & x_3 = 2h , \\
u_4 & = & u(x_4) & \text{at} & x_4 = 3h , \\
u_5 & = & u(x_5) & \text{at} & x_5 = 4h = 1 .
\end{array}
\tag{2.2}
$$

On the ith subinterval $x_i \le x \le x_{i+1}$, we will approximate the function $u(x)$ locally by the linear function

$$
\hat{u}_i(x) = \alpha_i + \beta_i x \qquad (i = 1, 2, 3, 4) ,
\tag{2.3}
$$

where α_i, β_i are the coefficients to be determined from the relations

$$
\hat{u}_i(x_i) = u_i \quad , \quad \hat{u}_i(x_{i+1}) = u_{i+1} .
\tag{2.4}
$$

From (2.3) and (2.4), it follows that

$$
\alpha_i = \frac{x_{i+1} u_i - x_i u_{i+1}}{x_{i+1} - x_i} ,
$$
$$
\beta_i = \frac{u_{i+1} - u_i}{x_{i+1} - x_i} \qquad (x_i \le x \le x_{i+1}) .
\tag{2.5}
$$

With these α_i, β_i, expression (2.3) becomes

$$
\hat{u}_i(x) = u_i \phi_i(x) + u_{i+1} \phi_{i+1}(x) ,
\tag{2.6}
$$

where the interpolations $\phi_i(x), \phi_{i+1}(x)$ have the form

$$
\phi_i(x) = \frac{x_{i+1} - x}{x_{i+1} - x_i} ,
$$
$$
\phi_{i+1}(x) = \frac{x - x_i}{x_{i+1} - x_i} \qquad (x_i \le x \le x_{i+1}) .
$$

The interpolation functions on the whole interval can also be expressed as follows.

$$
\phi_i(x) = \begin{cases} \frac{x - x_{i-1}}{x_i - x_{i-1}} & (x_{i-1} \le x \le x_i) \\ \frac{x_{i+1} - x}{x_{i+1} - x_i} & (x_i \le x \le x_{i+1}) \\ 0 & (\text{Otherwise}) . \end{cases}
\tag{2.7}
$$

We see that $\phi_i(x)$ resembles a "roof" with unit height, as illustrated in Figure 2.7(b). For this reason, the interpolation function given by (2.7) is often called

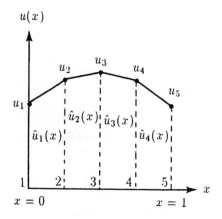

(a)   Approximating polygon.

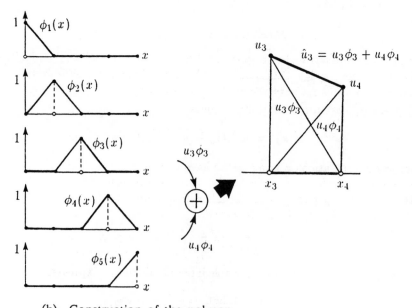

(b)   Construction of the polygon.

Figure 2.7: Interpolation using roof functions.

the *roof function*. We notice the following general properties of the finite element basis.

(i) $\phi_i(x_j) = \delta_{ij} = \begin{cases} 1 & (i = j) \\ 0 & (i \neq j) \end{cases}$, (2.8)

(ii) $\sum\limits_{i=1}^{n} \phi_i(x) = 1$.

The symbol δ_{ij} is called the Kronecker delta.

2.3   Method of Variations

We present an application of the Ritz method to the Couette flow described by the following boundary value problem: Suppose that we need to find the unknown velocity $u(y)$ in the x direction, satisfying the differential equation

$$\nu \frac{d^2 u}{dy^2} = 0 \ , \quad -a < y < a \qquad (2.9)$$

with a positive constant ν, subject to the boundary conditions

$$u = 0 \quad \text{at} \quad y = -a \ ,$$

and

$$u = U \quad \text{at} \quad y = a \ .$$

The problem is illustrated in Figure 2.8. We shall have to find a functional $F(u)$ to be minimized, which corresponds to the above boundary value problem. To this end, we consider a variation of $F(u)$, denoted by $\delta F(u)$, in the form

$$\delta F(u) = - \int_{-a}^{a} \nu \frac{d^2 u}{dy^2} \delta u \, dy = 0 \qquad (2.10)$$

for any variations δu of u with the constraint $\delta u = 0$ at the boundaries $y = \pm a$. The minus sign in front of the integral in (2.10) is added for convenience in order to transform the boundary value problem to a minimization problem.

Integration of (2.10) by parts yields

$$\begin{aligned} \delta F(u) &= [-\nu \frac{du}{dy} \delta u]_{-a}^{a} + \int_{-a}^{a} \nu \frac{du}{dy} \frac{d\delta u}{dy} \, dy \\ &= \int_{-a}^{a} \nu \frac{du}{dy} \frac{d\delta u}{dy} \, dy \ . \end{aligned} \qquad (2.11)$$

The last equality follows from the boundary constraints on δu. From the following differential formula of variations

$$\delta(\frac{du}{dy})^2 = 2 \frac{du}{dy} \frac{d\delta u}{dy} \ , \qquad (2.12)$$

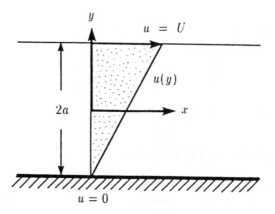

Figure 2.8: Couette flow problem.

equation (2.11) can be written in the form

$$\delta F(u) = \frac{1}{2} \int_{-a}^{a} \nu \, \delta(\frac{du}{dy})^2 \, dy = \delta \frac{1}{2} \int_{-a}^{a} \nu (\frac{du}{dy})^2 \, dy . \qquad (2.13)$$

This implies that the functional defined by

$$F(u) = \frac{1}{2} \int_{-a}^{a} \nu (\frac{du}{dy})^2 \, dy$$

is the one which we desired.

   We shall show that the functional attains a minimum value in the Couette flow problem. In fact, let $v(y)$ be any function satisfying the boundary conditions; $v(\pm a) = 0$. Put $v(y) = u(y) + \psi(y)$. Then we can see that $\psi(\pm a) = 0$. We know that

$$
\begin{aligned}
F(v) &= \frac{1}{2} \int_{-a}^{a} \nu (\frac{du}{dy} + \frac{d\psi}{dy})^2 \, dy \\
&= \frac{1}{2} \int_{-a}^{a} \nu (\frac{du}{dy})^2 \, dy + \int_{-a}^{a} \nu \frac{du}{dy} \frac{d\psi}{dy} \, dy \\
&\quad + \frac{1}{2} \int_{-a}^{a} \nu (\frac{d\psi}{dy})^2 \, dy = 0 .
\end{aligned}
$$

The last but one term on the right-hand side becomes

$$\int_{-a}^{a} \nu \frac{du}{dy} \frac{d\psi}{dy} \, dy = [\nu \frac{du}{dy} \psi]_{-a}^{a} - \int_{-a}^{a} \nu \frac{d^2 u}{dy^2} \psi \, dy = 0 ,$$

since u is the solution of the problem. Accordingly, we have

$$F(v) = F(u) + \frac{1}{2} \int_{-a}^{a} \nu (\frac{d\psi}{dy})^2 \, dy .$$

Since the second term on the right-hand side is always non-negative, we conclude

$$F(v) \geq F(u)$$

for any v. Here, the equality holds if and only if $v = u$.

We know that the boundary value problem given by (2.9) is equivalent to the following minimization problem: finding the function $u(y)$, which minimizes the functional

$$F(u) = \frac{1}{2} \int_{-a}^{a} \nu \left(\frac{du}{dy}\right)^2 dy \qquad (2.14)$$

among all functions satisfying the boundary conditions

$$u = 0 \quad \text{at} \quad y = -a$$

and

$$u = U \quad \text{at} \quad y = a .$$

2.3.1   Ritz finite element method.

We shall seek a finite element solution of the Couette flow problem, based on the minimization problem, using the Ritz method. To this end, we divide the whole interval $-a < y < a$ into four subintervals e_1, e_2, e_3, e_4 with equal length $h = a/2$, as shown in Figure 2.9, where the ith subinterval becomes the finite element $e_i = (y_i, y_{i+1})$.

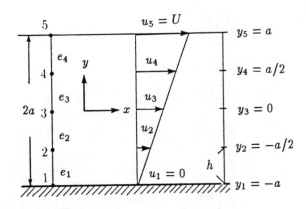

Figure 2.9: Finite element model in the Couette flow.

The unknown flow velocity $u(y)$ in the x direction is assumed to be approximated by

$$\hat{u}(y) = \sum_{i=1}^{5} u_i \, \phi_i(y) \qquad (2.15)$$

with the linear interpolation functions $\phi_i(y)$ given by (2.7). The functional F defined by (2.14), which is minimized by (2.15), can be evaluated as follows.

$$F(\hat{u}) = \frac{1}{2} \int_{-a}^{a} \nu (\frac{d\hat{u}}{dy})^2 \, dy = \sum_{i=1}^{4} \frac{1}{2} \int_{y_i}^{y_{i+1}} \nu (\frac{d\hat{u}}{dy})^2 \, dy .$$

On each subinterval e_i, the unknown velocity takes the approximate form

$$\hat{u}(y) = u_i \phi_i(y) + u_{i+1} \phi_{i+1}(y) .$$

With the notation

$$F_i(\hat{u}) = \frac{1}{2} \int_{y_i}^{y_{i+1}} \nu (\frac{d\hat{u}}{dy})^2 \, dy$$

$$= \frac{\nu}{2} \int_{e_i} (\frac{d\phi_i}{dy} u_i + \frac{d\phi_{i+1}}{dy} u_{i+1})^2 \, dy ,$$

we can write

$$F(\hat{u}) = \sum_{i=1}^{4} F_i(\hat{u}) .$$

The minimum of the approximated functional is found from the relationship

$$\frac{\partial F(\hat{u})}{\partial u_j} = \sum_{i=1}^{4} \frac{\partial F_i(\hat{u})}{\partial u_j} = 0 \qquad (j = 1, 2, 3, 4, 5) . \qquad (2.16)$$

Notice that

$$\frac{\partial F_i(\hat{u})}{\partial u_j} = 0 \qquad \text{for} \qquad j \neq i, i+1 .$$

It is only required to consider

$$\frac{\partial F_i(\hat{u})}{\partial u_i} = 0 \quad \text{and} \quad \frac{\partial F_i(\hat{u})}{\partial u_{i+1}} = 0 ,$$

such that

$$\nu \int_{e_i} (\frac{d\phi_i}{dy} u_i + \frac{d\phi_{i+1}}{dy} u_{i+1}) \frac{d\phi_i}{dy} \, dy = 0 , \qquad (2.17)$$

$$\nu \int_{e_i} (\frac{d\phi_i}{dy} u_i + \frac{d\phi_{i+1}}{dy} u_{i+1}) \frac{d\phi_{i+1}}{dy} \, dy = 0 . \qquad (2.18)$$

From (2.7), it is easy to see that

$$\int_{e_i} \frac{d\phi_i}{dy} \frac{d\phi_i}{dy} \, dy = \frac{1}{h} , \qquad (2.19)$$

$$\int_{e_i} \frac{d\phi_i}{dy} \frac{d\phi_{i+1}}{dy} \, dy = -\frac{1}{h} . \qquad (2.20)$$

Therefore, equations (2.17) and (2.18) result in the following system of element equations.

$$\left.\begin{array}{c} u_i - u_{i+1} = 0 \\ -u_i + u_{i+1} = 0 \end{array}\right\} . \tag{2.21}$$

In the matrix form, they can be written as

$$\begin{bmatrix} 1 & -1 \\ -1 & 1 \end{bmatrix} \begin{bmatrix} u_i \\ u_{i+1} \end{bmatrix} = \begin{bmatrix} 0 \\ 0 \end{bmatrix} . \tag{2.22}$$

We shall assemble the element equations in this matrix form according to relation (2.16). For this purpose, we extend expression (2.22) to the global matrix form as follows.

element e_1 :

$$\begin{bmatrix} 1 & -1 & 0 & 0 & 0 \\ -1 & 1 & 0 & 0 & 0 \\ 0 & 0 & 0 & 0 & 0 \\ 0 & 0 & 0 & 0 & 0 \\ 0 & 0 & 0 & 0 & 0 \end{bmatrix} \begin{bmatrix} u_1 \\ u_2 \\ u_3 \\ u_4 \\ u_5 \end{bmatrix} = \begin{bmatrix} 0 \\ 0 \\ 0 \\ 0 \\ 0 \end{bmatrix} ,$$

element e_2 :

$$\begin{bmatrix} 0 & 0 & 0 & 0 & 0 \\ 0 & 1 & -1 & 0 & 0 \\ 0 & -1 & 1 & 0 & 0 \\ 0 & 0 & 0 & 0 & 0 \\ 0 & 0 & 0 & 0 & 0 \end{bmatrix} \begin{bmatrix} u_1 \\ u_2 \\ u_3 \\ u_4 \\ u_5 \end{bmatrix} = \begin{bmatrix} 0 \\ 0 \\ 0 \\ 0 \\ 0 \end{bmatrix} ,$$

element e_3 :

$$\tag{2.23}$$

$$\begin{bmatrix} 0 & 0 & 0 & 0 & 0 \\ 0 & 0 & 0 & 0 & 0 \\ 0 & 0 & 1 & -1 & 0 \\ 0 & 0 & -1 & 1 & 0 \\ 0 & 0 & 0 & 0 & 0 \end{bmatrix} \begin{bmatrix} u_1 \\ u_2 \\ u_3 \\ u_4 \\ u_5 \end{bmatrix} = \begin{bmatrix} 0 \\ 0 \\ 0 \\ 0 \\ 0 \end{bmatrix} ,$$

element e_4 :

$$\begin{bmatrix} 0 & 0 & 0 & 0 & 0 \\ 0 & 0 & 0 & 0 & 0 \\ 0 & 0 & 0 & 0 & 0 \\ 0 & 0 & 0 & 1 & -1 \\ 0 & 0 & 0 & -1 & 1 \end{bmatrix} \begin{bmatrix} u_1 \\ u_2 \\ u_3 \\ u_4 \\ u_5 \end{bmatrix} = \begin{bmatrix} 0 \\ 0 \\ 0 \\ 0 \\ 0 \end{bmatrix} .$$

The assembly with respect to the element index i and the arrangement of these equations with respect to the nodal index j in expression (2.16) are accomplished through the summation of all element matrix equations. After assembling them, we have the following global matrix equation:

$$
\begin{bmatrix}
1 & -1 & 0 & 0 & 0 \\
-1 & 2 & -1 & 0 & 0 \\
0 & -1 & 2 & -1 & 0 \\
0 & 0 & -1 & 2 & -1 \\
0 & 0 & 0 & -1 & 1
\end{bmatrix}
\begin{bmatrix}
u_1 \\
u_2 \\
u_3 \\
u_4 \\
u_5
\end{bmatrix}
=
\begin{bmatrix}
0 \\
0 \\
0 \\
0 \\
0
\end{bmatrix}.
\tag{2.24}
$$

The coefficient matrix is called the *global matrix*. For the functional correspond-ing to the Couette flow, the matrix is symmetric but semi-positive definite. The semi-positiveness is seen from the fact that zero is an eigenvalue of the matrix corresponding to the eigenvector (1, 1, 1, 1, 1).

The boundary conditions are now substituted into (2.24). This means that $u_1 = 0$ and $u_5 = U$ are inserted into the unknown column vector. The number of unknowns will be reduced by two. It may be advantageous, however, in the computer implementation to keep the size and symmetry of the matrix. This results in

$$
\begin{bmatrix}
1 & 0 & 0 & 0 & 0 \\
0 & 2 & -1 & 0 & 0 \\
0 & -1 & 2 & -1 & 0 \\
0 & 0 & -1 & 2 & 0 \\
0 & 0 & 0 & 0 & 1
\end{bmatrix}
\begin{bmatrix}
u_1 \\
u_2 \\
u_3 \\
u_4 \\
u_5
\end{bmatrix}
=
\begin{bmatrix}
0 \\
0 \\
0 \\
U \\
U
\end{bmatrix}.
\tag{2.25}
$$

The coefficient matrix now becomes positive definite. This matrix equation can be written simply in the form

$$
[K]\{u\} = \{b\}.
\tag{2.26}
$$

The solution is given by

$$
\begin{bmatrix}
u_1 \\
u_2 \\
u_3 \\
u_4 \\
u_5
\end{bmatrix}
= U
\begin{bmatrix}
0 \\
1/4 \\
2/4 \\
3/4 \\
1
\end{bmatrix}.
\tag{2.27}
$$

The finite element solution expressed by (2.15) with the nodal values given by (2.27) coincides with the exact solution

$$
u(y) = \frac{U}{2}\left(1 + \frac{y}{a}\right).
$$

2.4   Method of Weighted Residuals

We shall apply the method of weighted residuals to the problem of the Couette flow. Suppose that $\hat{u}(y)$ is an approximate solution to equation (2.9). Due to the error incurred in the approximation we may have

$$
\nu \frac{d^2\hat{u}}{dy^2} \neq 0.
$$

With some arbitrary weighting functions $w(y)$, we now try to find $\hat{u}$ such that

$$\int_{-a}^{a} \nu \frac{d^2\hat{u}}{dy^2} w \, dy = 0 \ . \qquad (2.28)$$

The integration by parts reduces the order of differentiation for unknown $\hat{u}$, namely,

$$\left[\nu \frac{d\hat{u}}{dy} w \right]_{-a}^{a} - \int_{-a}^{a} \nu \frac{d\hat{u}}{dy} \frac{dw}{dy} dy = 0 \ .$$

The approximate solution for $\hat{u}(y)$ is assumed to have the form of (2.15). Since the values of $\hat{u}$ at $y = \pm a$ are prescribed by the boundary conditions, the weighting functions w are so chosen such that $w = 0$ at these two ends. This implies that

$$\int_{-a}^{a} \nu \frac{d\hat{u}}{dy} \frac{dw}{dy} dy = 0 \qquad (2.29)$$

for any w with $w = 0$ at $y = \pm a$.

2.4.1   Galerkin finite element method.

In the Galerkin finite element method, the base functions ϕ_i $(i = 1, 2, \ldots, n)$ are chosen as the weighting functions. After the global matrix equation has been obtained, the Dirichlet boundary condition is inserted into the equation.

We shall start with the substitution of the following expressions into (2.29).

$$\hat{u}(x) = \sum_{j=1}^{n} u_j \phi_j(y) \quad \text{and} \quad w(y) = \phi_i(y) \ , \qquad (2.30)$$

with $n = 5$, we obtain

$$\sum_{j=1}^{n} \int_{-a}^{a} \nu \frac{d\phi_i}{dy} \frac{d\phi_j}{dy} dy \, u_j = 0 \ . \qquad (2.31)$$

If we denote the integral by k_{ij}, we will have the following linear system of equations.

$$\sum_{j=1}^{n} k_{ij} u_j = 0 \qquad (i = 1, 2, \ldots n) \ . \qquad (2.32)$$

Substituting the boundary conditions; $u = 0$ at $y = -a$ and $u = U$ at $y = a$, we can obtain a global matrix equation of the form (2.26).

The finite element method is featured by its elementwise procedure, as we shall see in the following: from (2.29) we have

$$\int_{-a}^{a} \nu \frac{d\hat{u}}{dy} \frac{d\phi_j}{dy} dy = 0 \qquad (j = 1, 2, 3, 4, 5) \ .$$

The whole interval of this integration is divided into finite elements $e_i = (y_i, y_{i+1})$ as follows.

$$\sum_{i=1}^{4} \int_{y_i}^{y_{i+1}} \nu \frac{d\hat{u}}{dy} \frac{d\phi_j}{dy} \, dy = 0 \, .$$

On each element e_i, the approximate solution has the form

$$\hat{u}(y) = u_i \phi_i(y) + u_{i+1} \phi_{i+1}(y) \, .$$

Accordingly we shall write the equation on e_i as follows.

$$\int_{e_i} \nu \frac{d\hat{u}}{dy} \frac{d\phi_j}{dy} \, dy = 0 \qquad (j = i, i+1) \, .$$

The equations then become

$$\int_{e_i} \nu \frac{d\phi_i}{dy} \frac{d\phi_i}{dy} \, dy \, u_i + \int_{e_i} \nu \frac{d\phi_{i+1}}{dy} \frac{d\phi_i}{dy} \, dy \, u_{i+1} = 0 \, ,$$

and

$$\int_{e_i} \nu \frac{d\phi_i}{dy} \frac{d\phi_{i+1}}{dy} \, dy \, u_i + \int_{e_i} \nu \frac{d\phi_{i+1}}{dy} \frac{d\phi_{i+1}}{dy} \, dy \, u_{i+1} = 0 \, ,$$

respectively. Using (2.19) and (2.20), we obtain the same element equations as (2.21). After the assembly of all the element equations and insertion of the Dirichlet boundary conditions, we can obtain the solution given by (2.27).

We could anticipate from the preceding two examples for the Couette flow that the Ritz method is equivalent to the Galerkin method for the case when the boundary value problems have their functionals minimized.

2.4.2   Example in the Poiseuille flow.

We shall consider the differential equation

$$\mu \frac{d^2 u}{dy^2} + \alpha = 0 \quad , \qquad -a < y < a \qquad (2.33)$$

for unknown $u(y)$ with given constants α and μ, subject to the boundary conditions

$$u = 0 \quad \text{at} \quad y = \pm a \, .$$

The problem is illustrated in Figure 2.10. The exact solution is given by

$$u(y) = \frac{\alpha}{2\mu} (a^2 - y^2) \, , \qquad (2.34)$$

and the corresponding pressure p is determined by the relation

$$\frac{dp}{dx} = -\alpha \, .$$

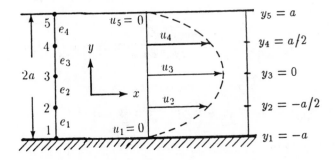

Figure 2.10: Finite element model in the Poiseuille flow.

In the Galerkin method, the unknown approximation $\hat{u}$ and weighting functions w are taken from the same function space consisting of all the linear combination of interpolation functions. We shall denote w by δu. We consider the following weighted residual expression corresponding to equation (2.33).

$$\int_{-a}^{a} (\, \mu \frac{d^2\hat{u}}{dy^2} + \alpha\,)\, \delta u\, dy \; = \; 0 \, ,$$

for any arbitrary δu, when $\delta u = 0$ at $y = \pm a$. Integration by parts yields

$$\int_{-a}^{a} \mu \frac{d\hat{u}}{dy} \frac{d\delta u}{dy}\, dy - \alpha \int_{-a}^{a} \delta u\, dy \; = \; 0 \, . \qquad (2.35)$$

For the element e_i, we shall write

$$\int_{y_i}^{y_{i+1}} \mu \frac{d\hat{u}_e}{dy} \frac{d\delta u_e}{dy}\, dy - \alpha \int_{y_i}^{y_{i+1}} \delta u_e\, dy \; = \; 0 \, , \qquad (2.36)$$

where the index e indicates the restriction of the functions $\hat{u}$ and δu to the element. Here we see

$$\hat{u}_e \; = \; u_i\phi_i + u_{i+1}\phi_{i+1} \quad , \quad \delta\hat{u}_e \; = \; \delta u_i\phi_i + \delta u_{i+1}\phi_{i+1} \, .$$

We notice that δu_i and δu_{i+1} inherit their arbitrariness from δu . Substituting these expressions into (2.36), we obtain

$$(\int_{e_i} \mu \frac{d\phi_i}{dy} \frac{d\phi_i}{dy} dy\, u_i + \int_{e_i} \mu \frac{d\phi_{i+1}}{dy} \frac{d\phi_i}{dy} dy\, u_{i+1} - \alpha \int_{e_i} \phi_i dy\,)\, \delta u_i$$

$$+ (\int_{e_i} \mu \frac{d\phi_i}{dy} \frac{d\phi_{i+1}}{dy} dy\, u_i + \int_{e_i} \mu \frac{d\phi_{i+1}}{dy} \frac{d\phi_{i+1}}{dy} dy\, u_{i+1} - \alpha \int_{e_i} \phi_{i+1} dy\,)\, \delta u_{i+1} \; = \; 0 \, .$$

It is important to realize that this equation is valid for any arbitrary δu_i and δu_{i+1}. This implies that each of the coefficients of δu_i and δu_{i+1} must be equal to zero. Namely,

$$\int_{e_i} \mu \frac{d\phi_i}{dy} \frac{d\phi_i}{dy} \, dy \, u_i + \int_{e_i} \mu \frac{d\phi_{i+1}}{dy} \frac{d\phi_i}{dy} \, dy \, u_{i+1} - \alpha \int_{e_i} \phi_i \, dy = 0,$$

and

$$\int_{e_i} \mu \frac{d\phi_i}{dy} \frac{d\phi_{i+1}}{dy} \, dy \, u_i + \int_{e_i} \mu \frac{d\phi_{i+1}}{dy} \frac{d\phi_{i+1}}{dy} \, dy \, u_{i+1} - \alpha \int_{e_i} \phi_{i+1} \, dy = 0.$$

It can be shown that

$$\int_{e_i} \phi_i \, dy = \int_{e_i} \phi_{i+1} \, dy = \frac{h}{2}.$$

Together with (2.19) and (2.20), we get the following element equations

$$\left. \begin{array}{r} \frac{\mu}{h} u_i - \frac{\mu}{h} u_{i+1} = \frac{h}{2}\alpha \\ -\frac{\mu}{h} u_i + \frac{\mu}{h} u_{i+1} = \frac{h}{2}\alpha \end{array} \right\}.$$

In matrix form, they become

$$\frac{\mu}{h} \begin{bmatrix} 1 & -1 \\ -1 & 1 \end{bmatrix} \begin{bmatrix} u_i \\ u_{i+1} \end{bmatrix} = \frac{h}{2}\alpha \begin{bmatrix} 1 \\ 1 \end{bmatrix}.$$

After assembly for all the finite elements, we obtain the following global matrix equation.

$$\frac{\mu}{h} \begin{bmatrix} 1 & -1 & 0 & 0 & 0 \\ -1 & 2 & -1 & 0 & 0 \\ 0 & -1 & 2 & -1 & 0 \\ 0 & 0 & -1 & 2 & -1 \\ 0 & 0 & 0 & -1 & 1 \end{bmatrix} \begin{bmatrix} u_1 \\ u_2 \\ u_3 \\ u_4 \\ u_5 \end{bmatrix} = \frac{h}{2}\alpha \begin{bmatrix} 1 \\ 2 \\ 2 \\ 2 \\ 1 \end{bmatrix}. \qquad (2.37)$$

When we apply the boundary conditions to this equation in such a way that the coefficient matrix with the same size remains symmetric, we obtain

$$\frac{\mu}{h} \begin{bmatrix} 1 & 0 & 0 & 0 & 0 \\ 0 & 2 & -1 & 0 & 0 \\ 0 & -1 & 2 & -1 & 0 \\ 0 & 0 & -1 & 2 & 0 \\ 0 & 0 & 0 & 0 & 1 \end{bmatrix} \begin{bmatrix} u_1 \\ u_2 \\ u_3 \\ u_4 \\ u_5 \end{bmatrix} = \frac{h}{2}\alpha \begin{bmatrix} 0 \\ 2 \\ 2 \\ 2 \\ 0 \end{bmatrix}. \qquad (2.38)$$

The solution is then given by

$$\begin{bmatrix} u_1 \\ u_2 \\ u_3 \\ u_4 \\ u_5 \end{bmatrix} = \alpha \frac{h^2}{\mu} \begin{bmatrix} 0 \\ 3/2 \\ 2 \\ 3/2 \\ 0 \end{bmatrix}. \qquad (2.39)$$

Table 2.1: Accuracy of the finite element solution.

Node	y	FEM	Exact
y_3	0.0	0.500	0.500
	0.1	0.475	0.495
	0.2	0.450	0.480
	0.3	0.425	0.455
	0.4	0.400	0.420
y_4	0.5	0.375	0.375
	0.6	0.300	0.320
	0.7	0.225	0.255
	0.8	0.150	0.180
	0.9	0.075	0.095
y_5	1.0	0.000	0.000

Table 2.1 shows the comparison of the finite element solution with the exact solution, when $\alpha = 1$, $h = 0.5$ and $\alpha/\mu = 1$. The approximate solution coincides with corresponding values of the exact solution only at the nodes, as shown in Figure 2.11.

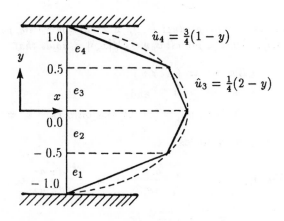

Figure 2.11: Approximate solution of Poiseuille flow.

2.4.3   Errors in One-dimensional Approximation

In this section we shall consider the boundary value problem

$$- \frac{d^2 u}{dx^2} = f(x) \qquad a < x < b , \tag{2.40}$$

$$u(a) = u(b) = 0 , \tag{2.41}$$

and we shall show a mathematical proof of convergence for the finite element approximate solutions to this problem.

In order to keep the mathematical rigour, we shall introduce some linear spaces of functions: We denote, by $C[a, b]$, the set of all real continuous functions defined on the closed interval $[a, b]$, with the sum of two functions and the product of the functions by real numbers. We denote, by $C^1[a, b]$ and $C^2[a, b]$ respectively, the sets of all functions which are once and twice continuously differentiable in $[a, b]$. Corresponding to the problem consisting of (2.40), (2.41), the following function space is introduced.

$$C_0^2[a, b] = \{ v \in C^2[a, b] \, / \, v(a) = v(b) = 0 \} .$$

The linear space $C[a, b]$ is characterised with the following scalar product and norms.

$$(u, v) = \int_a^b u(x) v(x) \, dx \qquad , \qquad \| u \|_2 = \sqrt{(u, v)} .$$

$$\| u \|_\infty = \max_{a \le x \le b} | u(x) | .$$

We assume that in (2.40) the right-hand side $f \in C[a, b]$. We denote the second-order differential operator by $\mathcal{L}$ and define it as

$$\mathcal{L} = - \frac{d^2}{dx^2} \; : \; C_0^2[a, b] \to C[a, b] .$$

Lemma 2.1 **With the scalar product** $(\cdot, \cdot)$, $\mathcal{L}$ **is a symmetric positive definite operator in** $C_0^2[a, b]$. **Namely, for all** $u, v \in C_0^2[a, b]$ **it holds that**

(i) $\qquad (\mathcal{L}u , v) = (u , \mathcal{L}v) ,$ \hfill (2.42)

(ii) $\qquad (\mathcal{L}u , u) \ge 0 , \quad$ and

$\qquad\qquad (\mathcal{L}u , u) = 0 \quad if \ and \ only \ if \quad u = 0 .$ \hfill (2.43)

Proof. Integration by parts yields

$$(\mathcal{L}u , v) = - \int_a^b \frac{d^2 u}{dx^2} v \, dx = \int_a^b \frac{du}{dx} \frac{dv}{dx} \, dx ,$$

$$(\mathcal{L}u , u) = \int_a^b \left(\frac{du}{dx}\right)^2 dx .$$

Symmetry indicated by (2.42) follows immediately from the first equalities. The positive definiteness indicated by (2.43) follows from the second equality; we see that $(\mathcal{L}u, u) = 0$ implies that $du/dx = 0$, which further implies, from the boundary condition (2.41), that $u = 0$. This completes the proof.

The bilinear form

$$(\mathcal{L}u, v) = \int_a^b \frac{du}{dx}\frac{dv}{dx}\,dx \qquad (2.44)$$

has been derived for u, v belonging to the space $C_0^2[a, b]$. We notice that the integration on the right-hand side of (2.44) makes sense if u and v are taken from the wider space $C_0^*[a, b]$ defined as follows.

(i) $C_0^2[a, b] \subset C_0^*[a, b] \subset C[a, b]$ with $v(a) = v(b) = 0$.

(ii) For any $v \in C_0^*[a, b]$, there exists an integer n depending on v such that $a = x_0 < x_1 < \ldots < x_n = b$ and $v \in C^1(x_{j-1}, x_j)$, $j = 1, 2, \ldots, n$.

(iii) The limits $v'(x_j \pm 0)$, $j = 1, 2, \ldots, n-1$ exist and are bounded.

The linear space $C_0^*[a, b]$ defined with the energy product given by (2.44) becomes a unitary space.

Lemma 2.2   In $C_0^*[a, b]$ the following inequalities hold.

$$\frac{1}{\alpha}\|v\|_\infty^2 \leq (\mathcal{L}v, v) \leq \alpha\|v'\|_\infty^2 \qquad (2.45)$$

with $\alpha = b - a$.

Proof. For any $v \in C_0^*[a, b]$, we can show that

$$v(x) = \int_a^x v'(t)\,dt \, .$$

From the Cauchy-Schwarz inequality, we have

$$v(x)^2 = \int_a^x dt \int_a^x |v'(t)|^2\,dt \leq (b-a)\int_a^b |v'(t)|^2\,dt \, ,$$

so that

$$\|v\|_\infty^2 \leq (b-a)\int_a^b |v'(t)|^2\,dt \, .$$

Hence

$$(\mathcal{L}v, v) = \int_a^b |v'(t)|^2\,dt \geq \frac{1}{b-a}\|v\|_\infty^2 \, .$$

The second inequality in (2.45) follows immediately from the above equality. This completes the proof.

We shall consider a functional, denoted by $F(v)$, on $C_0^*[a, b]$ by the definition

$$F(v) = (\mathcal{L}v, v) - 2(f, v) \, . \qquad (2.46)$$

Lemma 2.3   *Let u be a solution of (2.40) subject to (2.41). For any $v \in C_0^*[a, b]$, $v \neq u$, the following minimal property holds.*

$$F(v) > F(u) . \tag{2.47}$$

Proof. We see that

$$
\begin{aligned}
F(v) &= (\mathcal{L}v, v) - 2(\mathcal{L}u, v) + (\mathcal{L}u, u) - (\mathcal{L}u, u) \\
&= (\mathcal{L}(v - u), v - u) - (\mathcal{L}u, u) \\
&> -(\mathcal{L}u, u) = F(u) .
\end{aligned}
$$

This completes the proof.

Let the functions $\phi_1, \phi_2, \ldots, \phi_n$ be linearly independent in $C_0^*[a, b]$, and let S be the linearly spanned space of these n functions. We know that S is an n-dimensional subspace of $C_0^*[a, b]$. We put

$$v = \sum_{j=1}^{n} c_j \phi_j , \tag{2.48}$$

with the coefficients c_j of real numbers. From Lemma 2.3 it follows that

$$\min_{v \in S} F(v) = F(v^*) > F(u) , \qquad v^* \in S . \tag{2.49}$$

Theorem 2.1 For the exact solution u and the minimum v^*, it follows that

$$\text{(i)} \qquad (\mathcal{L}(v^* - u), v^* - u) = \min_{v \in S} (\mathcal{L}(v - u), v - u) , \tag{2.50}$$

$$\text{(ii)} \qquad \| v^* - u \|_{\infty} \leq \alpha \min_{v \in S} \| v' - u' \|_{\infty} . \tag{2.51}$$

Proof. Using the relation given by (2.49), we have

$$
\begin{aligned}
(\mathcal{L}(v - u), v - u) &= (\mathcal{L}v, v) - 2(\mathcal{L}u, v) + (\mathcal{L}u, u) \\
&= F(v) + (\mathcal{L}u, u) \\
&\geq F(v^*) + (\mathcal{L}u, u) \\
&= (\mathcal{L}(v^* - u), v^* - u) ,
\end{aligned}
$$

for any $v \in S$. This implies the equality of (2.50). To show (2.51) we use Lemma 2.2. It follows that

$$
\begin{aligned}
\frac{1}{\alpha} \| v^* - u \|_{\infty}^2 &\leq (\mathcal{L}(v^* - u, v^* - u) \\
&\leq (\mathcal{L}(v - u), v - u) \leq \alpha \| v' - u' \|_{\infty}^2 ,
\end{aligned}
$$

so that

$$\| v^* - u \|_{\infty}^2 \leq \alpha^2 \| v' - u' \|_{\infty}^2 .$$

This completes the proof.

We shall take the linear interpolation functions given by (2.7) as the base functions ϕ_j in S. From the result in Exercise 2.1, we know that

$$v(x) = \sum_{j=1}^{n} u(x_j)\phi_j(x), \qquad u \in C^2[a,b] \tag{2.52}$$

satisfies

$$\| v' - u' \|_\infty \leq \frac{h}{2} \| u'' \|_\infty \qquad \text{with} \qquad h = \max_j |x_j - x_{j-1}|. \tag{2.53}$$

Using this inequality, we can obtain an estimate of errors in the finite element solutions in terms of the mesh size h, as the next corollary reveals itself.

Corollary 2.1 The minimum solution v^* determined by (2.49) is identical with the finite element solution $\hat{u}$ and the error is estimated by

$$\| \hat{u} - u \|_\infty \leq \frac{\alpha}{2} \| f \|_\infty h. \tag{2.54}$$

See Exercise 2.2 for the proof. From this result, we know that the convergence is uniform and that the rate of the convergence is at least first order with respect to the mesh size h. We know also that the finite element method produces the exact solution if the right-hand side $f(x) = 0$, as we have seen in the examples for the Couette flow.

Exercises

2.1 Show that, if $u(x) \in C^2[a, b]$, then the linearly interpolated function v given by (2.52) satisfies the error estimates

$$| u'(x) - v'(x) | \leq \frac{h}{2} \max_{a \leq x \leq b} | u''(x) |,$$

$$| u(x) - v(x) | \leq \frac{h^2}{4} \max_{a \leq x \leq b} | u''(x) |.$$

Hint : In the subinterval $x_i \leq x \leq x_{i+1}$, we have

$$v'(x) = \frac{u(x_{i+1}) - u(x_i)}{x_{i+1} - x_i},$$

from which we have

$$u'(x) - v'(x) = u'(x) - \frac{u(x_{i+1}) - u(x) - \{ u(x_i) - u(x) \}}{x_{i+1} - x_i}.$$

By the Taylor's expansion theorem, we have

$$u(x_{i+1}) - u(x) = (x_{i+1} - x) u'(x) + \frac{1}{2} (x_{i+1} - x)^2 u''(\xi_1),$$

and

$$u(x_i) - u(x) = (x_i - x) u'(x) + \frac{1}{2} (x_i - x)^2 u''(\xi_2),$$

with $x_i \leq \xi_1$, $\xi_2 \leq x_{i+1}$. Hence, we can see that

$$| u'(x) - v'(x) | \leq \frac{1}{2} \frac{(x_{i+1} - x)^2 + (x_i - x)^2}{x_{i+1} - x_i} \max_{a \leq x \leq b} | u''(x) |$$

$$\leq \frac{1}{2} \frac{(x_{i+1} - x_i)^2}{x_{i+1} - x_i} \max_{a \leq x \leq b} | u''(x) |.$$

Moreover, we notice that

$$u(x) - v(x) = \begin{cases} \int_{x_i}^{x} \{ u'(t) - v'(t) \} dt, & x_i \leq x \leq \frac{x_i + x_{i+1}}{2} \\ \int_{x_{i+1}}^{x} \{ u'(t) - v'(t) \} dt, & \frac{x_i + x_{i+1}}{2} \leq x \leq x_{i+1} \end{cases}.$$

Therefore, we obtain

$$| u(x) - v(x) | \leq \max_{a \leq x \leq b} | u'(x) - v'(x) | \frac{h}{2}.$$

Incidentally as a remark, we can show that

$$\| u' - v' \|_2 \leq h \| u'' \|_2.$$

2.2 Prove that the minimum solution v^* determined by (2.49) is identical with the Galerkin finite element solution $\hat{u}$.

Hint : The Galerkin finite element solution assumes the form

$$\hat{u} \;=\; \sum_{j=1}^{n} c_j\,\phi_j$$

with the roof functions ϕ_j given by (2.7), where c_j are unknowns to be determined by the following linear system of equations.

$$(\,\mathcal{L}\hat{u}\,,\,\phi_i\,) \;=\; (\,f\,,\,\phi_i\,)\,, \qquad i = 1, 2, \ldots, n.$$

To be more specific, they read

$$\sum_{j=1}^{n}(\,\mathcal{L}\phi_j\,,\,\phi_i\,)\,c_j \;=\; (\,f\,,\,\phi_i\,)\,.$$

On the other hand, the functional (2.46) with v given by (2.48) has the expressions

$$
\begin{aligned}
F(v) \;&=\; (\,\mathcal{L}v\,,\,v\,) - 2\,(\,f\,,\,v\,)\\[4pt]
&=\; \sum_{j=1}^{n}\sum_{i=1}^{n}(\,\mathcal{L}\phi_j\,,\,\phi_i\,)\,c_j\,c_i \;-\; 2\sum_{i=1}^{n}(\,f\,,\,\phi_i\,)\,c_i\,.
\end{aligned}
$$

The necessary condition for the minimum;

$$\frac{\partial F(v)}{\partial c_i} \;=\; 0\,, \qquad i = 1, 2, \ldots, n$$

will give the same system of linear equations for unknown c_j.

2.3 Prove the error estimate:

$$\|\,\hat{u}' - u'\,\|_2 \;\le\; \frac{\sqrt{\alpha}}{2}\,h^2\,\|\,f\,\|_\infty$$

for the finite element solution $\hat{u}$ given in Exercise 2.2.

Hint From (2.5) with $v^* = \hat{u}$, we have

$$(\,\mathcal{L}(\hat{u}-u)\,,\,\hat{u}-u\,) \;\le\; (\,\mathcal{L}(v-u)\,,\,v-u\,)\,, \qquad v \in \mathcal{S}$$

because the minimum solution is identical with the finite element solution. Let v be the function given by (2.52), which interpolates u. Both sides of the above inequality become

$$
\begin{aligned}
\|\,\hat{u}' - u'\,\|_2^2 \;&\le\; \int_a^b \Big|\,\frac{d}{dx}(v-u)\,\Big|^2\,dx\\[4pt]
&\le\; (b-a)\,\|\,v'-u'\,\|_\infty^2 \;\le\; \frac{\alpha}{4}\,h^2\,\|\,u''\,\|_\infty^2\,.
\end{aligned}
$$

The last inequality followed from (2.53).

2.4 (Nitsche's trick) Prove the error estimate:

$$\| \hat{u} - u \|_2 \leq \frac{\sqrt{\alpha}}{2} h^2 \| f \|_\infty .$$

Hint: The exact solution u and the finite element solution $\hat{u}$ satisfy the equalities:

$$(\mathcal{L} u , v) = (f , v) , \qquad \forall v \in C_0^*[a, b]$$

and

$$(\mathcal{L} \hat{u} , \hat{v}) = (f , \hat{v}) , \qquad \forall \hat{v} \in \mathcal{S} ,$$

respectively. In particular it follows that

$$(\mathcal{L} u , \hat{v}) = (f , \hat{v}) .$$

By subtracting the last two equations, we obtain

$$(\mathcal{L} (\hat{u} - u), \hat{v}) = 0 .$$

We consider a solution $\psi(x)$ of the boundary value problem:

$$\mathcal{L} \psi = \hat{u}(x) - u(x) , \qquad a < x < b$$
$$\psi(a) = \psi(b) = 0 .$$

The solution ψ satisfies

$$(\mathcal{L} \psi , v) = (\hat{u} - u , v) , \qquad \forall v \in C_0^*[a, b] .$$

Put $v = \hat{u} - u$. Then we get

$$
\begin{aligned}
(\hat{u} - u , \hat{u} - u) &= (\mathcal{L} \psi , \hat{u} - u) \\
&= (\psi , \mathcal{L} (\hat{u} - u)) \\
&= (\psi - \hat{v} , \mathcal{L} (\hat{u} - u)) .
\end{aligned}
$$

Let ψ^I denote the interpolate of ψ:

$$\psi^I(x) = \sum_{j=1}^{n} \psi(x_j) \phi_j(x) \in \mathcal{S}$$

and put $\hat{v} = \psi^I$. Then, it follows from the Cauchy-Schwarz inequality that

$$
\begin{aligned}
\| \hat{u} - u \|_2^2 &= \int_a^b \frac{d}{dx} (\psi - \psi^I) \frac{d}{dx} (\hat{u} - u) \, dx \\
&\leq \| \psi' - \psi^I{}' \|_2 \, \| \hat{u}' - u' \|_2 .
\end{aligned}
$$

From the remark to Exercise 2.1 and Exercise 2.3 we see that

$$
\begin{aligned}
\| \hat{u} - u \|_2^2 &\leq h \| \psi'' \|_2 \frac{\sqrt{\alpha}}{2} h \| f \|_\infty \\
&= \frac{\sqrt{\alpha}}{2} h^2 \| \hat{u} - u \|_2 \| f \|_\infty .
\end{aligned}
$$

2.5 The $n \times n$ matrix

$$[A] = \begin{bmatrix} a_{11} & a_{12} & \cdots & a_{1n} \\ a_{21} & a_{22} & \cdots & a_{2n} \\ \vdots & \vdots & & \vdots \\ a_{n1} & a_{n2} & \cdots & a_{nn} \end{bmatrix}$$

is said to be (*strictly*)*diagonally dominant*, if the following inequalities hold.

$$\sum_{\substack{j=1 \\ j \neq i}}^{n} \mid a_{ij} \mid \, < \mid a_{ii} \mid, \qquad (i = 1, 2, \ldots, n).$$

Show that the diagonally dominant matrix is invertible. Moreover, the matrix is said to be *irreducibly diagonally dominant*, if the *digraph* corresponding to the matrix $[A]$ is *strongly connected* and the following inequalities hold

$$\sum_{\substack{j=1 \\ j \neq i}}^{n} \mid a_{ij} \mid \, \leq \mid a_{ii} \mid, \qquad (i = 1, 2, \ldots, n)$$

where at least one inequality must hold with the unequal sign. It can be proved that the irreducibly diagonally dominant matrix is invertible.

Hint : Suppose that a strictly diagonally dominant matrix $[A]$ is not invertible. Since the matrix is singular, the homogeneous system of linear equations

$$\sum_{j=1}^{n} a_{ij} x_j = 0 \; , \quad i = 1, 2, \ldots, n$$

has a non-trivial solution; $(x_1^*, x_2^*, \ldots, x_n^*) \neq (0, 0, \ldots, 0)$. Let

$$\mid x_i^* \mid = \max_{1 \leq j \leq n} \mid x_j^* \mid \, .$$

Then we can see that

$$\sum_{j=1}^{n} a_{ij} x_j^* = 0 \; , \quad a_{ii} x_i^* = -\sum_{\substack{j=1 \\ j \neq i}}^{n} a_{ij} x_j^* \, ,$$

so that

$$\mid a_{ii} \mid \, \leq \sum_{\substack{j=1 \\ j \neq i}}^{n} \mid a_{ij} \mid \frac{\mid x_j^* \mid}{\mid x_i^* \mid} \leq \sum_{\substack{j=1 \\ j \neq i}}^{n} \mid a_{ij} \mid \, < \mid a_{ii} \mid,$$

which leads to a contradiction.

2.6 Consider an elastic rod over $a \leq x \leq b$ with a uniform cross section $A(m^2)$. Two ends of the rod are assumed to be firmly fixed. Due to some external body force $f\,(N/m^3)$ in the x direction, suppose that the rod is deformed in that direction. Let $u\,(m)$ denote the displacement. The strain $\epsilon\,(-)$ and stress $\sigma\,(Pa)$ are defined by

$$\epsilon = \frac{du}{dx} \quad , \quad \sigma = E\,\epsilon$$

with the Young's modulus $E\,(Pa)$. The equilibrium of forces can be expressed by the equation

$$\frac{d\sigma}{dx} + f = 0 \; .$$

The *total energy* produced in the rod is given by

$$\Pi(u) = \int_a^b \frac{1}{2} E\,\epsilon^2\,A\,dx - \int_a^b f\,u\,A\,dx \; ,$$

where the first term on the right-hand side is called the *strain energy* and the second is called the *potential energy*. Show that

$$\Pi(u) = \frac{A}{2}\,F(u)$$

with the functional given by (2.46).

Chapter 3

TWO-DIMENSIONAL PROBLEMS

In this chapter we shall describe the finite element method as applied to the boundary value problem of the Laplace equation in two dimensions. Many flow problems of interest are governed by the two-dimensional Laplace equation. The Laplace equation will lead through its introduction to more complicated applications of the finite element method. Linear triangular finite elements are exclusively discussed herein. This chapter ends with a few mathematical arguments about the accuracy and convergence of the finite element approximation.

3.1 Laplace Equation in Two Dimensions

Let Ω be a domain in the x, y-plane, whose boundary is denoted by Γ. We consider the following mixed boundary value problem: find the unknown velocity potential Φ satisfying the Laplace equation

$$\nabla^2 \Phi = 0 \quad \text{in} \quad \Omega \,, \tag{3.1}$$

subject to the boundary conditions

$$\Phi = \Phi_B \quad \text{on} \quad \Gamma_\Phi \,,$$

and

$$\frac{\partial \Phi}{\partial n} = V_n \quad \text{on} \quad \Gamma_n \,, \tag{3.2}$$

where values of the velocity potential Φ_B are prescribed on the part of the boundary Γ_Φ, and the flow velocity V_n in the direction of the external normal n is prescribed on the rest of the boundary Γ_n.

3.2   Element Subdivision

3.2.1   Triangulation

The domain Ω is subdivided into a finite number of small triangles. In general, straight sided triangular elements can not accurately represent curved boundaries, inevitably there will be small gaps between the straight sided element and the curved boundary. We shall ignore these gaps for the sake of simplicity in our proceeding discussion. To reduce computational errors caused by the approximation of the geometry and by the interpolation of unknown functions, it is recommended to follow the guidelines set below in the course of the triangulation:

(i) *Use fine mesh in the vicinity of abrupt changes in the geometry, boundary conditions, and material properties.*
(ii) *Avoid extremely irregular mesh. Don't use obtuse triangles.*
(iii) *Triangles should not overlap each other.*
(iv) *No area gap is allowed in the internal subdivision.*
(v) *Nodes are located only at the vertices of the triangles. Diagonals intersect at the corners of the domain.*
(vi) *If the physical problem has some sort of symmetry, subdivide the domain according to the symmetry.*

Figure 3.1 illustrates some remarks on these guidelines. The reader should keep in mind again that the subdivision dominates the ultimate quality of the finite element solution.

For potential flow problems, the triangulation is relatively easy. However, the domains of the more complicated problems must be carefully subdivided in order to increase the accuracy of their numerical solutions, as we will see in subsequent chapters: For viscous flow problems, great care must be taken about the subdivision. Spurious phenomena and numerical divergence may often completely deteriorate the finite element solution, when the domain, particularly the areas right behind the obstacles or near the open boundaries, is improperly subdivided.

3.2.2   Numbering of elements and nodes

The elements and nodes in the triangulation are numbered consecutively. In this book, element numbers are circled. Any jump in the numbering is not permitted.

The total coefficient matrix obtained from the two-dimensional Laplace equation often has the form

$$[K] = \begin{bmatrix} k_{11} & k_{12} & k_{13} & k_{14} & & & & 0 \\ k_{21} & k_{22} & k_{23} & k_{24} & k_{25} & & & \\ k_{31} & k_{32} & & & & & k_{n-3\ n} \\ k_{41} & k_{42} & & & & & k_{n-2\ n} \\ & k_{52} & & & & & k_{n-1\ n} \\ 0 & & k_{n\ n-3} & k_{n\ n-2} & k_{n\ n-1} & & k_{n\ n} \end{bmatrix}.$$

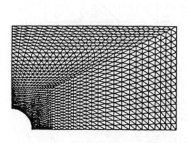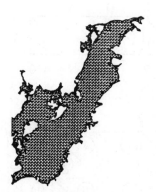

(i)   Mesh grading.

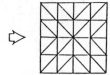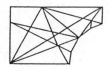

(ii)   Avoid extremely
irregular mesh.

(iii)   Overlapping is
unacceptable.

A gap appears
by element shrink.

(iv)   Gap inside the domain is unacceptable.

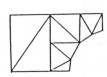

(v)   A node located on
an element side is
unacceptable.

(vi)   Symmetrical subdivision
according to the
physical symmetry.

Figure 3.1: Some remarks on the element subdivision.

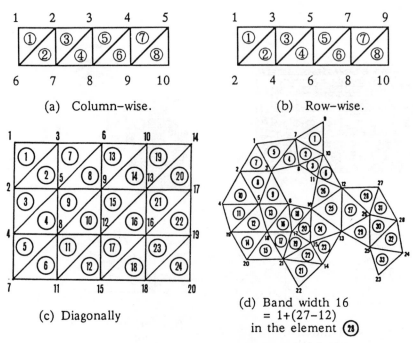

(a) Column-wise.   (b) Row-wise.

(c) Diagonally

(d) Band width 16
    = 1+(27−12)
    in the element ㉘

Figure 3.2: Node numberings.

The matrix is sparse and the non-zero entities cluster around the main diagonal.
A matrix of this form is called a *band matrix*, because the non-zero entities are
banded. Owing to the symmetry of the coefficient matrix, only the diagonal and
upper-diagonal non-zero entities can be stored in the computer in the following
compact form.

$$\begin{bmatrix} k_{11} & k_{12} & k_{13} & k_{14} \\ & k_{22} & k_{23} & k_{24} & k_{25} \\ & & \cdot & \cdot & \cdot & \cdot \\ & & & \cdot & \cdot & \cdot & \cdot \\ & & & & \cdot & k_{n-3\ n} \\ & & & k_{n-2\ n} \\ & & k_{n-1\ n} \\ k_{n\ n} \end{bmatrix} \cdot \qquad (3.3)$$

In this way, the computer storage is saved when one omits the superfluous
zeros. The size of the computer storage is reduced, if the difference between
node numbers of two adjacent nodes is minimized. The number of columns in the
matrix formation described by (3.3) is called the *band width*. The band width is
equal to one plus the maximum among all the differences between node numbers
of two adjacent nodes. The band width will change as the nodes are numbered
in different order. Minimizing the band width is also useful in reducing the
computing time, particularly when the corresponding linear system of equations

is solved by the method of band matrix Gaussian elimination.

We consider two different numbering of nodes for the meshes, as shown in Figure 3.2. We shall observe how the corresponding band widths will change. Corresponding to Figure 3.2(a) and (b), the total matrices are produced in the form

$$
\begin{bmatrix}
* & * & * & * & * & * & & & & 0 \\
* & * & * & * & * & * & * & & & \\
* & * & * & * & * & * & * & * & & \\
* & * & * & * & * & * & * & * & * & \\
* & * & * & * & * & * & * & * & * & * \\
* & * & * & * & * & * & * & * & * & * \\
& * & * & * & * & * & * & * & * & * \\
& & * & * & * & * & * & * & * & * \\
& & & * & * & * & * & * & * & * \\
0 & & & & * & * & * & * & * & *
\end{bmatrix}
\text{ and }
\begin{bmatrix}
* & * & * & & & & & & & 0 \\
* & * & * & * & & & & & & \\
* & * & * & * & * & & & & & \\
& * & * & * & * & * & & & & \\
& & * & * & * & * & * & & & \\
& & & * & * & * & * & * & & \\
& & & & * & * & * & * & * & \\
& & & & & * & * & * & * & \\
0 & & & & & & * & * & *
\end{bmatrix}
$$

respectively. These band widths are 6 and 3.

The band width is reduced for a slender domain, if the nodes are numbered in the direction parallel to the shorter side. For domains of more complex geometry, the optimal numbering which results in the minimum band width is intuitively difficult. A sample computer program for *renumbering* in the optimal order is included in the attached diskette.

3.2.3   Topological data

All nodal coordinates and the set of three nodal numbers associated with each triangular finite element constitute the topological data. In Figure 3.3, the nodal coordinates are read in the cartesian coordinate system (x, y). The order of the nodal numbers belonging to an element are read counterclockwise. They are listed as i_1, i_2, i_3 in Table 3.1.

The generation of the topological data is time consuming. As the geometry becomes more complex, the preparation for the topological input data is error-prone. A computer program, which generates the mesh automatically, is included in the attached diskette.

3.3   Two-dimensional Linear Interpolation

Let e denote a general triangular element, as shown in Figure 3.4. Let i_1, i_2, i_3 be the associated three nodal numbers, which are always read in the counter-clockwise rotation. Moreover, let $(x_1, y_1), (x_2, y_2), (x_3, y_3)$ be their coordinates.

We consider a linear function u^e in x and y, interpolating a smooth function $u(x, y)$ with the nodal values u_1, u_2, u_3 in the element e. The linear function takes the form

$$u^e(x, y) = \alpha_1 + \alpha_2 x + \alpha_3 y \tag{3.4}$$

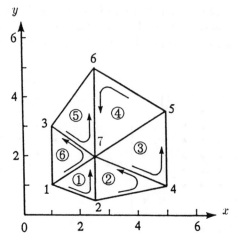

Figure 3.3: Topology constituent in the finite elements.

Table 3.1: Topological data.

Node number	(x, y)
1	(1.0 , 1.0)
2	(2.5 , 0.5)
3	(1.0 , 3.0)
4	(5.0 , 1.0)
5	(5.0 , 3.5)
6	(2.5 , 5.0)
7	(2.5 , 2.0)

(a) Nodal coordinates

Element number	Nodes		
	i_1	i_2	i_3
①	1	2	7
②	2	4	7
③	7	4	5
④	5	6	7
⑤	3	7	6
⑥	1	7	3

(b) Element connectivity

with the coefficients $\alpha_1, \alpha_2, \alpha_3$ to be determined from the relations

$$
\begin{aligned}
u_1 &= \alpha_1 + \alpha_2 x_1 + \alpha_3 y_1 & \text{at} \quad & i_1, \\
u_2 &= \alpha_1 + \alpha_2 x_2 + \alpha_3 y_2 & \text{at} \quad & i_2, \\
u_3 &= \alpha_1 + \alpha_2 x_3 + \alpha_3 y_3 & \text{at} \quad & i_3.
\end{aligned}
\tag{3.5}
$$

They render a linear system of equations to unknown $\alpha_1, \alpha_2, \alpha_3$, namely

$$
\begin{bmatrix} 1 & x_1 & y_1 \\ 1 & x_2 & y_2 \\ 1 & x_3 & y_3 \end{bmatrix}
\begin{bmatrix} \alpha_1 \\ \alpha_2 \\ \alpha_3 \end{bmatrix}
=
\begin{bmatrix} u_1 \\ u_2 \\ u_3 \end{bmatrix}.
\tag{3.6}
$$

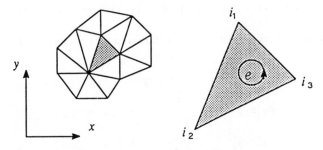

Figure 3.4: A triangular element.

The inverse of the coefficient matrix is given by

$$
\begin{bmatrix} 1 & x_1 & y_1 \\ 1 & x_2 & y_2 \\ 1 & x_3 & y_3 \end{bmatrix}^{-1} = \frac{1}{2\,\Delta^e} \begin{bmatrix} a_1 & a_2 & a_3 \\ b_1 & b_2 & b_3 \\ c_1 & c_2 & c_3 \end{bmatrix} , \tag{3.7}
$$

where

$$
\begin{aligned}
a_1 = x_2\,y_3 - x_3\,y_2 , & \quad b_1 = y_2 - y_3 , & \quad c_1 = x_3 - x_2 , \\
a_2 = x_3\,y_1 - x_1\,y_3 , & \quad b_2 = y_3 - y_1 , & \quad c_2 = x_1 - x_3 , \\
a_3 = x_1\,y_2 - x_2\,y_1 , & \quad b_3 = y_1 - y_2 , & \quad c_3 = x_2 - x_1 ,
\end{aligned} \tag{3.8}
$$

and

$$
\begin{aligned}
\Delta^e &= \text{the area of the triangle } e \\
&= \frac{1}{2} \begin{vmatrix} 1 & x_1 & y_1 \\ 1 & x_2 & y_2 \\ 1 & x_3 & y_3 \end{vmatrix} = b_2\,c_3 - b_3\,c_2 .
\end{aligned} \tag{3.9}
$$

The solution $(\alpha_1, \alpha_2, \alpha_3)$ of (3.6) is therefore given by

$$
\begin{bmatrix} \alpha_1 \\ \alpha_2 \\ \alpha_3 \end{bmatrix} = \frac{1}{2\,\Delta^e} \begin{bmatrix} a_1 & a_2 & a_3 \\ b_1 & b_2 & b_3 \\ c_1 & c_2 & c_3 \end{bmatrix} \begin{bmatrix} u_1 \\ u_2 \\ u_3 \end{bmatrix} .
$$

We know that the linear function expressed by (3.4) is given as follows.

$$
\begin{aligned}
u^e(x, y) &= \frac{1}{2\,\Delta^e} (a_1 + b_1\,x + c_1\,y)\,u_1 \\
&+ \frac{1}{2\,\Delta^e} (a_2 + b_2\,x + c_2\,y)\,u_2 \\
&+ \frac{1}{2\,\Delta^e} (a_3 + b_3\,x + c_3\,y)\,u_3 .
\end{aligned} \tag{3.10}
$$

We shall express this in the form

$$u^e(x, y) = \phi_1(x, y) u_1 + \phi_2(x, y) u_2 + \phi_3(x, y) u_3 \qquad (3.11)$$

using the following *linear interpolation functions* for the triangular element.

$$\phi_1(x, y) = \frac{1}{2 \Delta^e} (a_1 + b_1 x + c_1 y),$$

$$\phi_2(x, y) = \frac{1}{2 \Delta^e} (a_2 + b_2 x + c_2 y), \qquad (3.12)$$

$$\phi_3(x, y) = \frac{1}{2 \Delta^e} (a_3 + b_3 x + c_3 y).$$

The linear combination given by (3.11) is geometrically interpreted, as illustrated in Figure 3.5. In the same way as in (2.8) for the one-dimensional interpolation functions, the two-dimensional interpolation functions here satisfy the following relationships,

(i) $\phi_i(x_j, y_j) = \delta_{ij}$,

(ii) $\phi_1(x, y) + \phi_2(x, y) + \phi_3(x, y) \equiv 1$.

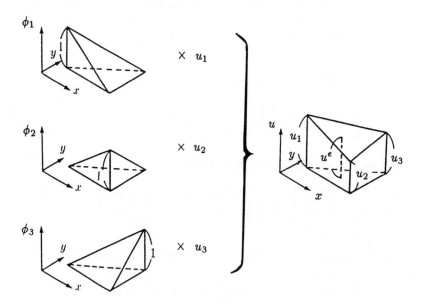

Figure 3.5: Linear triangular interpolation functions.

3.4 Gauss-Green's Formula

With arbitrary weighting functions $\delta\Phi(x,y)$, we consider the following weighted residual form of the Laplace equation

$$\iint_\Omega (\nabla^2\Phi)\,\delta\Phi\,dx\,dy \;=\; 0\;. \tag{3.13}$$

For the domain presented in Figure 3.6, we shall transform the double integral in (3.13) into corresponding iterated integrals. Referring to Figure 3.6 (a) , we see that

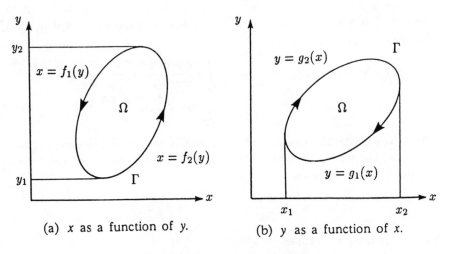

(a) x as a function of y.   (b) y as a function of x.

Figure 3.6: Integration paths.

$$\iint_\Omega \frac{\partial^2\Phi}{\partial x^2}\,\delta\Phi\,dx\,dy \;=\; \int_{y_1}^{y_2} dy \int_{f_1(y)}^{f_2(y)} \frac{\partial^2\Phi}{\partial x^2}\,\delta\Phi\,dx$$

$$=\; \int_{y_1}^{y_2} \frac{\partial\Phi}{\partial x}(f_2(y),y)\,\delta\Phi(f_2(y),y)\,dy$$

$$-\int_{y_1}^{y_2} \frac{\partial\Phi}{\partial x}(f_1(y),y)\,\delta\Phi(f_1(y),y)\,dy \;-\; \int_{y_1}^{y_2} dy \int_{f_1(y)}^{f_2(y)} \frac{\partial\Phi}{\partial x}\frac{\partial\delta\Phi}{\partial x}\,dx$$

$$=\; \int_{y_1}^{y_2} \frac{\partial\Phi}{\partial x}(f_2(y),y)\,\delta\Phi(f_2(y),y)\,dy$$

$$+\int_{y_2}^{y_1} \frac{\partial\Phi}{\partial x}(f_1(y),y)\,\delta\Phi(f_1(y),y)\,dy \;-\; \iint_\Omega \frac{\partial\Phi}{\partial x}\frac{\partial\delta\Phi}{\partial x}\,dx\,dy$$

$$=\; \oint_\Gamma \frac{\partial\Phi}{\partial x}\,\delta\Phi\,dy \;-\; \iint_\Omega \frac{\partial\Phi}{\partial x}\frac{\partial\delta\Phi}{\partial x}\,dx\,dy\;, \tag{3.14}$$

where the notation $\oint_\Gamma$ indicates the contour integral in counterclockwise direction along the closed curve Γ. Referring to Figure 3.6 (b), we can show that

$$\iint_\Omega \frac{\partial^2 \Phi}{\partial y^2} \delta\Phi \, dx \, dy \;=\; \int_{x_1}^{x_2} dx \int_{g_1(x)}^{g_2(x)} \frac{\partial^2 \Phi}{\partial y^2} \delta\Phi \, dy$$

$$=\; \int_{x_1}^{x_2} \frac{\partial \Phi}{\partial y}(x, g_2(x)) \, \delta\Phi\,(x, g_2(x)) \, dx$$

$$-\int_{x_1}^{x_2} \frac{\partial \Phi}{\partial y}(x, g_1(x)) \, \delta\Phi\,(x, g_1(x)) \, dx \;-\; \iint_\Omega \frac{\partial \Phi}{\partial y} \frac{\partial \delta\Phi}{\partial y} \, dx \, dy$$

$$=\; -\oint_\Gamma \frac{\partial \Phi}{\partial y} \delta\Phi \, dx \;-\; \iint_\Omega \frac{\partial \Phi}{\partial y} \frac{\partial \delta\Phi}{\partial y} \, dx \, dy\;. \qquad (3.15)$$

The minus sign before the contour integral comes out from the fact that the rotation of Γ is taken clockwise. Combining these two results derived in (3.14) and (3.15), we have

$$\iint_\Omega (\nabla^2 \Phi) \delta\Phi \, dx \, dy \;=\; -\iint_\Omega \left(\frac{\partial \Phi}{\partial x} \frac{\partial \delta\Phi}{\partial x} + \frac{\partial \Phi}{\partial y} \frac{\partial \delta\Phi}{\partial y} \right) dx \, dy$$

$$+\oint_\Gamma \delta\Phi \left(\frac{\partial \Phi}{\partial x} dy - \frac{\partial \Phi}{\partial y} dx \right). \qquad (3.16)$$

We now consider the physical interpretation of the contour integral in (3.16). Recall that the velocity components u and v are calculated from the velocity potential Φ by the formulas

$$u \;=\; \frac{\partial \Phi}{\partial x}\;, \qquad v \;=\; \frac{\partial \Phi}{\partial y}\;.$$

The contour integral becomes

$$\oint_\Gamma \delta\Phi \left(\frac{\partial \Phi}{\partial x} dy - \frac{\partial \Phi}{\partial y} dx \right) \;=\; \oint_\Gamma \delta\Phi\,(u\,dy - v\,dx)\;.$$

Let $d\Gamma$ denote the infinitesimal line element of Γ. Let α be the angle between $d\Gamma$ and the x axis, and let β be the angle between the vector (u, v) and the x axis, as illustrated in Figure 3.7. From geometrical considerations we obtain

$$dx \;=\; d\Gamma \cos\alpha\;, \qquad dy \;=\; d\Gamma \sin\alpha$$

and

$$u \;=\; V \cos\beta\;, \qquad v \;=\; V \sin\beta$$

where V is the magnitude of the flow vector. It follows that

$$u\,dy - v\,dx \;=\; V\,(\cos\beta \sin\alpha - \sin\beta \cos\alpha)\,d\Gamma$$

$$=\; V \sin(\alpha - \beta)\,d\Gamma\;.$$

The last expression refers to the normal component of the flow vector on the boundary:

$$V_n \;=\; \frac{\partial \Phi}{\partial n}\;.$$

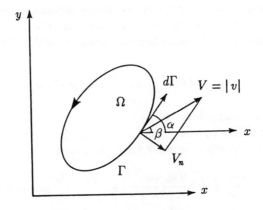

Figure 3.7: Angles α, β.

Therefore,

$$\oint_\Gamma \delta\Phi\,(u\,dy - v\,dx) = \oint_\Gamma \delta\Phi\,\frac{\partial\Phi}{\partial n}\,d\Gamma\ .$$

The equality

$$\iint_\Omega (\nabla^2\Phi)\delta\Phi\,dx\,dy = \oint_\Gamma \frac{\partial\Phi}{\partial n}\,\delta\Phi\,d\Gamma - \iint_\Omega \nabla\Phi\cdot\nabla\delta\Phi\,dx\,dy \qquad (3.17)$$

is known as the *Gauss–Green's formula*. From (3.13) we have

$$\iint_\Omega \left(\frac{\partial\Phi}{\partial x}\frac{\partial\delta\Phi}{\partial x} + \frac{\partial\Phi}{\partial y}\frac{\partial\delta\Phi}{\partial y}\right)dx\,dy - \oint_\Gamma \frac{\partial\Phi}{\partial n}\,\delta\Phi\,d\Gamma = 0\ . \qquad (3.18)$$

This is called the *weak form*, corresponding to the Laplace equation. From the boundary condition (3.2), the value of Φ on part of the boundary Γ_Φ is known. The weighting functions $\delta\Phi$ can be arbitrary, but $\delta\Phi = 0$ along the boundary Γ_Φ. On the rest of the boundary Γ_n, the normal velocity V_n is specified, according to (3.2).

3.5   Galerkin Finite Element Method

The weak form (3.18) is the starting point of the finite element approximation procedure. The domain Ω is subdivided into a series of triangular finite elements. The unknown velocity potential Φ is approximated linearly on each element e in the form

$$\begin{aligned}
\Phi(x,y) &= \phi_1(x,y)\,\Phi_1 + \phi_2(x,y)\,\Phi_2 + \phi_3(x,y)\,\Phi_3 \\
&= \sum_{\alpha=1}^{3} \phi_\alpha(x,y)\,\Phi_\alpha\ , \qquad (3.19)
\end{aligned}$$

where ϕ_α are the finite element interpolation functions given by (3.12), and Φ_α are the nodal values. In the Galerkin method, the weighting function $\delta\Phi$ is assumed to belong to the same function space, to which Φ belongs. Accordingly we may take

$$\delta\Phi(x,y) \; = \; \sum_{\alpha=1}^{3} \phi_\alpha(x,y)\,\delta\Phi_\alpha \tag{3.20}$$

with arbitrary nodal values $\delta\Phi_\alpha$.

We shall confine the weak form (3.18) to an element e. On substituting (3.19) and (3.20) into the confined weak form, we obtain

$$\sum_{\alpha=1}^{3}\{\sum_{\beta=1}^{3}\int\!\!\int_e (\frac{\partial\phi_\alpha}{\partial x}\frac{\partial\phi_\beta}{\partial x} \; + \; \frac{\partial\phi_\alpha}{\partial y}\frac{\partial\phi_\beta}{\partial y})\,dx\,dy\,\Phi_\beta$$

$$- \int_{\Gamma_n^e} \frac{\partial\Phi}{\partial n}\phi_\alpha\,d\Gamma\}\,\delta\Phi_\alpha \; = \; 0\,,$$

where Γ_n^e denote the intersection, if any, of Γ_n and the sides of the triangle e. From the arbitrariness of $\delta\Phi_\alpha$ ($\alpha = 1, 2, 3$), it follows that

$$\sum_{\beta=1}^{3}\int\!\!\int_e (\frac{\partial\phi_\alpha}{\partial x}\frac{\partial\phi_\beta}{\partial x} \; + \; \frac{\partial\phi_\alpha}{\partial y}\frac{\partial\phi_\beta}{\partial y})\,dx\,dy\,\Phi_\beta$$

$$- \int_{\Gamma_n^e} V_n\,\phi_\alpha\,d\Gamma \; = \; 0 \quad (\alpha = 1, 2, 3)\,. \tag{3.21}$$

Denoting the coefficients

$$K_{\alpha\beta}^e \; = \; \int\!\!\int_e (\frac{\partial\phi_\alpha}{\partial x}\frac{\partial\phi_\beta}{\partial x} \; + \; \frac{\partial\phi_\alpha}{\partial y}\frac{\partial\phi_\beta}{\partial y})\,dx\,dy\,, \tag{3.22}$$

and

$$F_\alpha^e \; = \; \int_{\Gamma_n^e} V_n\,\phi_\alpha\,d\Gamma\,, \tag{3.23}$$

then the element equations (3.21) can be written as

$$\sum_{\beta=1}^{3} K_{\alpha\beta}^e\,\Phi_\beta \; = \; F_\alpha^e \quad (\alpha = 1, 2, 3)\,. \tag{3.24}$$

In the matrix form, they become

$$\left[\begin{array}{c} K_{\alpha\beta}^e \end{array}\right]\left[\begin{array}{c} \Phi_1 \\ \Phi_2 \\ \Phi_3 \end{array}\right] = \left[\begin{array}{c} F_1^e \\ F_2^e \\ F_3^e \end{array}\right].$$

From (3.12), we can see that the entities of the coefficient matrix are specified as follows.

$$[K^e] \; = \; \frac{1}{4\,\Delta^e}\left[\begin{array}{ccc} b_1^{\;2}+c_1^{\;2} & b_1 b_2 + c_1 c_2 & b_1 b_3 + c_1 c_3 \\ & b_2^{\;2}+c_2^{\;2} & b_2 b_3 + c_2 c_3 \\ \text{symmetric} & & b_3^{\;2}+c_3^{\;2} \end{array}\right]. \tag{3.25}$$

To carry out the integration given by (3.23), we assume that the element e has the side 2-3 on the boundary Γ_n, as illustrated in Figure 3.8. Let (n, s) be the local coordinate system with the normal and tangential directions given by n and s respectively at node 2. The transformation between the global coordinate system (x, y) and the local one is given by

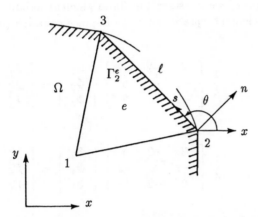

Figure 3.8: A finite element neighbouring on the boundary.

$$\begin{bmatrix} x \\ y \end{bmatrix} = -\frac{1}{\ell} \begin{bmatrix} b_1 & -c_1 \\ c_1 & b_1 \end{bmatrix} \begin{bmatrix} n \\ s \end{bmatrix} - \begin{bmatrix} x_2 \\ y_2 \end{bmatrix} , \qquad (3.26)$$

where

$$\begin{aligned} b_1 &= y_2 - y_3 = -\ell \sin \theta , \\ c_1 &= x_3 - x_2 = \ell \cos \theta \end{aligned}$$

with the angle θ between the side 2-3 and the x axis.

The boundary is expressed by $n = 0$. From (3.26) we have

$$\begin{aligned} \begin{bmatrix} F_1^e \\ F_2^e \\ F_3^e \end{bmatrix} &= \frac{1}{2\Delta^e} \int_0^\ell \begin{bmatrix} a_1 + b_1 x + c_1 y \\ a_2 + b_2 x + c_2 y \\ a_3 + b_3 x + c_3 y \end{bmatrix} V_n \, ds \\ &= \ell V_n \begin{bmatrix} 0 \\ 1/2 \\ 1/2 \end{bmatrix} . \end{aligned} \qquad (3.27)$$

This result can be interpreted in the following way: The total flux in the n-direction through the element boundary Γ_n^e is ℓV_n. The total flux is shared by the two boundary nodes. The contribution of the boundary flux to the element equations is therefore given by (3.27).

3.6   Total Equations

The element equations

$$[K^e]\{\Phi\} = \{F^e\} \tag{3.28}$$

are obtained for all elements e. The next step is to assemble all of these equations. We shall illustrate the assembly process with the next example.

For this purpose, we consider the finite element subdivision as shown in Figure 3.9. The element equations for the Laplace equation (3.1) are given by

element ①

$$\frac{1}{4} \begin{bmatrix} 4 & -2 & -2 \\ -2 & 2 & 0 \\ -2 & 0 & 2 \end{bmatrix} \begin{bmatrix} \Phi_1 \\ \Phi_2 \\ \Phi_3 \end{bmatrix} = \begin{bmatrix} F_1^1 \\ F_2^1 \\ F_3^1 \end{bmatrix},$$

element ②

$$\frac{1}{4} \begin{bmatrix} 2 & -2 & 0 \\ -2 & 4 & -2 \\ 0 & -2 & 2 \end{bmatrix} \begin{bmatrix} \Phi_2 \\ \Phi_4 \\ \Phi_3 \end{bmatrix} = \begin{bmatrix} F_2^2 \\ F_4^2 \\ F_3^2 \end{bmatrix},$$

element ③

$$\frac{1}{4} \begin{bmatrix} 4 & -2 & -2 \\ -2 & 2 & 0 \\ -2 & 0 & 2 \end{bmatrix} \begin{bmatrix} \Phi_3 \\ \Phi_4 \\ \Phi_5 \end{bmatrix} = \begin{bmatrix} F_3^3 \\ F_4^3 \\ F_5^3 \end{bmatrix},$$

element ④

$$\frac{1}{4} \begin{bmatrix} 2 & -2 & 0 \\ -2 & 4 & -2 \\ 0 & -2 & 2 \end{bmatrix} \begin{bmatrix} \Phi_4 \\ \Phi_6 \\ \Phi_5 \end{bmatrix} = \begin{bmatrix} F_4^4 \\ F_6^4 \\ F_5^4 \end{bmatrix}.$$

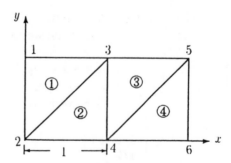

Figure 3.9: A 4-element model.

To assist the understanding, we rearrange these element equations in the global matrix form as follows.

element ①

$$\frac{1}{4}\begin{bmatrix} 4 & -2 & -2 & 0 & 0 & 0 \\ -2 & 2 & 0 & 0 & 0 & 0 \\ -2 & 0 & 2 & 0 & 0 & 0 \\ 0 & 0 & 0 & 0 & 0 & 0 \\ 0 & 0 & 0 & 0 & 0 & 0 \\ 0 & 0 & 0 & 0 & 0 & 0 \end{bmatrix} \begin{bmatrix} \Phi_1 \\ \Phi_2 \\ \Phi_3 \\ \Phi_4 \\ \Phi_5 \\ \Phi_6 \end{bmatrix} = \begin{bmatrix} F_1^1 \\ F_2^1 \\ F_3^1 \\ 0 \\ 0 \\ 0 \end{bmatrix},$$

element ②

$$\frac{1}{4}\begin{bmatrix} 0 & 0 & 0 & 0 & 0 & 0 \\ 0 & 2 & 0 & -2 & 0 & 0 \\ 0 & 0 & 2 & -2 & 0 & 0 \\ 0 & -2 & -2 & 4 & 0 & 0 \\ 0 & 0 & 0 & 0 & 0 & 0 \\ 0 & 0 & 0 & 0 & 0 & 0 \end{bmatrix} \begin{bmatrix} \Phi_1 \\ \Phi_2 \\ \Phi_3 \\ \Phi_4 \\ \Phi_5 \\ \Phi_6 \end{bmatrix} = \begin{bmatrix} 0 \\ F_2^2 \\ F_3^2 \\ F_4^2 \\ 0 \\ 0 \end{bmatrix},$$

element ③

$$\frac{1}{4}\begin{bmatrix} 0 & 0 & 0 & 0 & 0 & 0 \\ 0 & 0 & 0 & 0 & 0 & 0 \\ 0 & 0 & 4 & -2 & -2 & 0 \\ 0 & 0 & -2 & 2 & 0 & 0 \\ 0 & 0 & -2 & 0 & 2 & 0 \\ 0 & 0 & 0 & 0 & 0 & 0 \end{bmatrix} \begin{bmatrix} \Phi_1 \\ \Phi_2 \\ \Phi_3 \\ \Phi_4 \\ \Phi_5 \\ \Phi_6 \end{bmatrix} = \begin{bmatrix} 0 \\ 0 \\ F_3^3 \\ F_4^3 \\ F_5^3 \\ 0 \end{bmatrix},$$

element ④

$$\frac{1}{4}\begin{bmatrix} 0 & 0 & 0 & 0 & 0 & 0 \\ 0 & 0 & 0 & 0 & 0 & 0 \\ 0 & 0 & 0 & 0 & 0 & 0 \\ 0 & 0 & 0 & 2 & 0 & -2 \\ 0 & 0 & 0 & 0 & 2 & -2 \\ 0 & 0 & 0 & -2 & -2 & 4 \end{bmatrix} \begin{bmatrix} \Phi_1 \\ \Phi_2 \\ \Phi_3 \\ \Phi_4 \\ \Phi_5 \\ \Phi_6 \end{bmatrix} = \begin{bmatrix} 0 \\ 0 \\ 0 \\ F_4^4 \\ F_5^4 \\ F_6^4 \end{bmatrix}.$$

Taking the sum of these coefficient matrices and the sum of the corresponding right-hand sides, we obtain the following total equations,

$$\frac{1}{4}\begin{bmatrix} 4 & -2 & -2 & 0 & 0 & 0 \\ -2 & 4 & 0 & -2 & 0 & 0 \\ -2 & 0 & 8 & -4 & -2 & 0 \\ 0 & -2 & -4 & 8 & 0 & -2 \\ 0 & 0 & -2 & 0 & 4 & -2 \\ 0 & 0 & 0 & -2 & -2 & 4 \end{bmatrix} \begin{bmatrix} \Phi_1 \\ \Phi_2 \\ \Phi_3 \\ \Phi_4 \\ \Phi_5 \\ \Phi_6 \end{bmatrix} = \begin{bmatrix} F_1^1 \\ F_2^1 + F_2^2 \\ F_3^1 + F_3^2 + F_3^3 \\ F_4^2 + F_4^3 + F_4^4 \\ F_5^3 + F_5^4 \\ F_6^4 \end{bmatrix}. \quad (3.29)$$

We write the total equation in the form

$$[K]\{\Phi\} = \{F\}. \quad (3.30)$$

The coefficient matrix $[K]$ is symmetric but is singular. We insert the Dirichlet boundary condition $\Phi = \Phi_B$ into (3.30) at all nodes on Γ_Φ so that the resultant total equation has the form

$$[K_B]\{\Phi\} = \{F_B\} \tag{3.31}$$

with the symmetric coefficient matrix $[K_B]$. Now, the new coefficient matrix $[K_B]$ has become positive definite.

3.7   Errors in Two-dimensional Approximation

We shall consider the Dirichlet problem: Find unknown $u(x, y)$ satisfying the Poisson equation

$$-\Delta u = f(x, y) \quad \text{in} \quad \Omega \tag{3.32}$$

subject to the boundary condition

$$u = 0 \quad \text{on} \quad \Gamma = \partial\Omega . \tag{3.33}$$

Here we assume that Ω is a bounded domain enclosed by a finite number of smooth simple curves. We shall show an error estimate and the mathematical proof of convergence for the finite element approximate solution (3.11).

We denote the closure of Ω by $\bar{\Omega}$. Let $C_0(\bar{\Omega})$ denote the set of all continuous functions defined on $\bar{\Omega}$, satisfying (3.33). We introduce the linear space $C_0^2(\bar{\Omega})$ by the definition $C_0^2(\bar{\Omega}) = C^2(\bar{\Omega}) \cap C_0(\bar{\Omega})$, i.e., the set of all twice-continuously differentiable functions on $\bar{\Omega}$, that can be extended outside $\bar{\Omega}$ in C^2-class, with the property given by (3.33). The space $C_0^2(\bar{\Omega})$ is defined with the following scalar product and its norm.

$$(u, v) = \int_\Omega u(x, y) v(x, y) \, d\Omega \quad , \quad \|u\|_2 = \sqrt{(u, u)} ,$$

where $d\Omega = dxdy$. We assume that the right-hand side $f \in C(\bar{\Omega})$ in (3.32). Then we recast the Dirichlet problem as

$$\mathcal{L} = -\Delta : \quad C_0^2(\bar{\Omega}) \to C(\bar{\Omega}) .$$

Lemma 3.1 **With the scalar product $(\cdot, \cdot)$, $\mathcal{L}$ is a symmetric positive definite operator in $C_0^2(\bar{\Omega})$.**
Proof. From the Gauss-Green's formula (3.17), we see that

$$(\mathcal{L}u, v) = -\int_\Omega (\Delta u) v \, d\Omega$$

$$= -\int_\Gamma \frac{\partial u}{\partial n} v \, d\Gamma + \int_\Omega \nabla u \cdot \nabla v \, d\Omega ,$$

for any $u, v \in C_0^2(\bar{\Omega})$. Since $v = 0$ on Γ, we have a bilinear form given by

$$(\mathcal{L}u, v) = \int_\Omega \nabla u \cdot \nabla v \, d\Omega . \tag{3.34}$$

The symmetry follows immediately from this equality. Moreover, we can see that $(\mathcal{L}u, u) \geq 0$. The equality $(\mathcal{L}u, u) = 0$ implies that

$$\frac{\partial u}{\partial x} = \frac{\partial u}{\partial y} = 0 \quad \text{in} \quad \Omega ,$$

which implies further that $u = \text{const.} = 0$ from the boundary condition. This completes the proof.

We notice that the bilinear form (3.34) makes sense even if u and v are taken from the wider space $C_0^*(\bar{\Omega})$ defined as follows:

(i) $C_0^1(\bar{\Omega}) \subset C_0^*(\bar{\Omega}) \subset C_0(\bar{\Omega})$,

(ii) For any $v \in C_0^*(\bar{\Omega})$, there exists a finite partition of $\bar{\Omega}$; $\bar{\Omega} = \cup_{j=1}^n \bar{\Omega}_j$, $\Omega_j \cap \Omega_k = \phi \, (j \neq k)$, depending on v such that $v \in C^1(\bar{\Omega}_j), j = 1, 2, \ldots, n$.

The linear space $C_0^*(\bar{\Omega})$ with the *energy product* (3.34) becomes a unitary space. We denote $\|\nabla v\|_2 = \sqrt{(\mathcal{L}v, v)}$.

Lemma 3.2 (Friedrichs inequality) With the diameter d of the set Ω, the following inequality holds

$$\|v\|_2 \leq d\|\nabla v\|_2 \tag{3.35}$$

for all $v \in C_0^*(\bar{\Omega})$.
Proof. Since Ω is bounded with the diameter d, we can assume that

$$K = \{(x, y) \, / \, 0 \leq x \leq d, 0 \leq y \leq d\} \supset \Omega$$

without loss of generality. We consider an extension $v^* \in C_0^*(K)$ of $v \in C_0^*(\bar{\Omega})$, by the definition

$$v^* = \begin{cases} v & \text{on} \quad \bar{\Omega} \\ 0 & \text{on} \quad K \setminus \bar{\Omega} . \end{cases}$$

It holds that

$$v^*(x, y) = \int_0^x \frac{\partial v^*}{\partial s}(s, y) \, ds .$$

Using the Schwarz inequality, we have

$$v^*(x,y)^2 \leq \int_0^x ds \int_0^x |\frac{\partial v^*}{\partial s}(s, y)|^2 \, dx \, dy \leq d \int_0^d |\frac{\partial v^*}{\partial s}(s, y)|^2 \, ds .$$

Therefore we see that

$$\|v\|_2^2 = \int_\Omega |v(x, y)|^2 \, dx \, dy = \int_K |v^*(x, y)|^2 \, dx \, dy$$

$$\leq d \int_K \int_0^d |\frac{\partial v^*}{\partial s}(s, y)|^2 \, ds \, dx \, dy = d^2 \int_K |\frac{\partial v^*}{\partial x}(x, y)|^2 \, dx \, dy$$

$$\leq d^2 \int_K (|\frac{\partial v^*}{\partial x}|^2 + |\frac{\partial v^*}{\partial y}|^2) \, dx \, dy = d^2 \int_\Omega |\nabla v|^2 \, d\Omega .$$

This completes the proof.

We introduce a new norm $\|v\|_w$ called the *Sobolev norm* in $C_0^*(\bar{\Omega})$, by the definition

$$\|v\|_w^2 = \|v\|_2^2 + \|\nabla v\|_2^2 . \tag{3.36}$$

Lemma 3.3 In $C_0^*(\bar{\Omega})$ the following inequalities hold

$$\frac{1}{\alpha}\|v\|_w^2 \le (\mathcal{L}v, v) \le \alpha\|\nabla v\|_2^2 \tag{3.37}$$

with a positive constant α.

Proof. Using the Friedrichs inequality (3.35), we have

$$\|v\|_w^2 \le (1+d^2)\|\nabla v\|_2^2 = (1+d^2)(\mathcal{L}v, v) .$$

We put $\alpha = 1 + d^2$. The validity of the second inequality in (3.37) is obvious. This completes the proof.

We introduce a functional $F(v)$ defined on $C_0^*(\bar{\Omega})$, by the definition

$$F(v) = (\mathcal{L}v, v) - 2(f, v) . \tag{3.38}$$

Lemma 3.4 Let u be a solution of the Dirichlet problem given by (3.32) and (3.33). For any $v \in C_0^*(\bar{\Omega})$, $v \ne u$, the following minimal property holds:

$$F(v) > F(u) . \tag{3.39}$$

The proof is rendered to the reader as an exercise.

Let $\phi_1, \phi_2, \ldots, \phi_n$ be linearly independent functions in $C_0^*(\bar{\Omega})$, and let S be the linearly spanned space of those n functions. We know that S is an n-dimensional subspace of $C_0^*(\bar{\Omega})$ and so we can write

$$v = \sum_{j=1}^{n} c_j \phi_j . \tag{3.40}$$

From Lemma 3.4 we see that

$$\min_{v \in S} F(v) = F(v^*) > F(u) , \quad v^* \in S . \tag{3.41}$$

Theorem 3.1 For the exact solution u and the minimum v^*, it follows that

$$\text{(i)} \quad (\mathcal{L}(v^* - u), v^* - u) = \min_{v \in S}(\mathcal{L}(v - u), v - u) , \tag{3.42}$$

$$\text{(ii)} \quad \|v^* - u\|_w \le \alpha \min_{v \in S}\|\nabla(v - u)\|_2 . \tag{3.43}$$

For the sake of simplicity, we assume that Ω is a polygonal convex domain. This domain is subdivided into triangular finite elements. Among all the triangles, let h denote the largest side length and let θ be the smallest angle. We shall take the linear interpolation functions (3.12) as the basis in $\mathcal{S}$, in which the approximate solutions are sought. Due to Zlámal [1968] we have

Proposition 3.1 Assume that $u \in C^2(\bar{\Omega})$ with the second derivatives bounded by M on $\bar{\Omega}$, i.e.,

$$\max_{\Omega} \{ |\frac{\partial^2 u}{\partial x^2}| , |\frac{\partial^2 u}{\partial x \partial y}| , |\frac{\partial^2 u}{\partial y^2}| \} \leq M .$$

Then the function u^I obtained from u by the interpolation using the linear triangular finite elements satisfies the inequality

$$\max_{\Omega} \{ |\frac{\partial u^I}{\partial x} - \frac{\partial u}{\partial x}| , |\frac{\partial u^I}{\partial y} - \frac{\partial u}{\partial y}| \} \leq \frac{6M}{\sin \theta} h . \tag{3.44}$$

The finite element method is used to find u_h in $\mathcal{S}$, such that

$$(\mathcal{L}u_h , v) = (f , v) \tag{3.45}$$

for any $v \in \mathcal{S}$. Using Proposition 3.1, we obtain

Corollary 3.1 The minimum solution v^* determined by (3.41) coincides with the finite element solution u_h. The error in u_h is estimated using the Sobolev norm by

$$\| u_h - u \|_w \leq \frac{\sqrt{72 \, |\Omega| \, M \, \alpha}}{\sin \theta} h \tag{3.46}$$

with the area $|\Omega|$ of the domain.
Proof. For the second part of the assertions, we notice that the piecewise linearly interpolated function u^I belongs to $\mathcal{S}$. From Proposition 3.1 we can see that

$$\| \nabla(u^I - u) \|_2^2 = \int_\Omega | \nabla(u^I - u) |^2 \, d\Omega$$

$$= \int_\Omega \{ |\frac{\partial u^I}{\partial x} - \frac{\partial u}{\partial x}|^2 + |\frac{\partial u^I}{\partial y} - \frac{\partial u}{\partial y}|^2 \} d\Omega$$

$$\leq 2(\frac{6M}{\sin \theta} h)^2 \int_\Omega d\Omega .$$

Therefore, from (3.43) we have (3.46). This completes the proof.

Exercises

3.1 Verify (3.7)-(3.9).

3.2 Verify the element equations in Section 3.6.

3.3 Insert the following boundary conditions into (3.29).

$$\Phi = 6 \quad \text{on} \quad x = 0 \; , \quad \Phi = 4 \quad \text{on} \quad x = 2 \; ,$$

$$\frac{\partial \Phi}{\partial y} = 0 \quad \text{on} \quad y = 0 \quad \text{and} \quad y = 1 \; .$$

Solve the resultant total equation corresponding to (3.31) and compare the finite element solution with the exact analytical solution.

Hint: The total equation, after the above boundary conditions are inserted, is given by

$$\begin{bmatrix} 1 & 0 & 0 & 0 & 0 & 0 \\ 0 & 1 & 0 & 0 & 0 & 0 \\ 0 & 0 & 2 & -1 & 0 & 0 \\ 0 & 0 & -1 & 2 & 0 & 0 \\ 0 & 0 & 0 & 0 & 1 & 0 \\ 0 & 0 & 0 & 0 & 0 & 1 \end{bmatrix} \begin{bmatrix} \Phi_1 \\ \Phi_2 \\ \Phi_3 \\ \Phi_4 \\ \Phi_5 \\ \Phi_6 \end{bmatrix} = \begin{bmatrix} 6 \\ 6 \\ 5 \\ 5 \\ 4 \\ 4 \end{bmatrix} \; ,$$

The exact solution is given by $\Phi(x,y) = 6 - x$.

3.4 Consider the *Friedrichs-Keller finite element mesh* with $h = 1/2$, as shown in Figure 3.10. Using this mesh, derive the total equation corresponding to (3.30) for the Laplace equation (3.1). Compare the total equation with the equation obtained from conventional five-point finite difference approximation, for example,

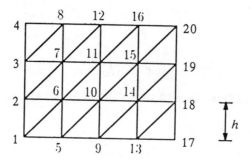

Figure 3.10: Friedrichs-Keller mesh.

$$\frac{1}{h^2} \left(4\Phi_6 - \Phi_2 - \Phi_5 - \Phi_7 - \Phi_{10} \right) = 0 \; .$$

3.5 Prove Lemma 3.4 and Theorem 3.1.

Chapter 4

POTENTIAL FLOWS

In this chapter we present some numerical examples of the potential flow in two dimensions. These examples will serve as good exercises for the implementation of the finite element method using a PC. The flows are governed by the Laplace equation. An inviscid flow around a circular cylinder is modelled in terms of either streamfunction or velocity potential. The groundwater flow regarded as a potential flow is also presented. Singularities caused by corners and abrupt changes in the boundary conditions are discussed.

4.1 Basic Equations in Potential Flows

We consider two-dimensional flows of an ideal fluid occupying the domain Ω in the rectangular coordinate system (x, y). If the flow is irrotational, the velocity components u and v are expressed in terms of the velocity potential Φ as follows.

$$u = \frac{\partial \Phi}{\partial x} \quad , \quad v = \frac{\partial \Phi}{\partial y} . \tag{4.1}$$

We assume that the fluid is incompressible, then the equation of continuity becomes

$$\frac{\partial u}{\partial x} + \frac{\partial v}{\partial y} = 0 . \tag{4.2}$$

Substitution of (4.1) into this equation results in the Laplace equation

$$\frac{\partial^2 \Phi}{\partial x^2} + \frac{\partial^2 \Phi}{\partial y^2} = 0 . \tag{4.3}$$

The boundary Γ of the flow domain consists of two parts: On one part of the boundary, the value of the velocity potential Φ_B is prescribed. On the rest of the boundary, the normal component of the velocity V_n is prescribed;

$$\Phi = \Phi_B \quad \text{on} \quad \Gamma_\Phi ,$$

$$\frac{\partial \Phi}{\partial n} = V_n \quad \text{on} \quad \Gamma_n . \tag{4.4}$$

On the other hand, the velocity components can also be derived from the streamfunction ψ as follows.

$$u = \frac{\partial \psi}{\partial y} \quad , \quad v = -\frac{\partial \psi}{\partial x} \, . \tag{4.5}$$

Since the flow is assumed to be irrotational, the vorticity ω is equal to zero, $i.e.$,

$$\omega = \frac{\partial v}{\partial x} - \frac{\partial u}{\partial y} = 0 \, . \tag{4.6}$$

Substitution of (4.5) into (4.6) again results in the Laplace equation

$$\frac{\partial^2 \psi}{\partial x^2} + \frac{\partial^2 \psi}{\partial y^2} = 0 \, . \tag{4.7}$$

Once again, the boundary is assumed to consist of two parts: On part of the boundary, the value of the streamfunction ψ_B is prescribed. On the rest of the boundary, the tangential component of the velocity V_s is prescribed;

$$\psi = \psi_B \quad \text{on} \quad \Gamma_\psi \, ,$$

$$\frac{\partial \psi}{\partial n} = -V_s \quad \text{on} \quad \Gamma_s \, . \tag{4.8}$$

Thus, the potential flow can be formulated in terms of either the velocity potential or the streamfunction.

4.2   Discretisation of the Weak Form

We consider the discretisation of the weak form corresponding to the Laplace equation using the finite element method. Let $\delta\psi$ be weighting functions, being arbitrary but $\delta\psi = 0$ on Γ_ψ, where the value of ψ is prescribed. To derive the weak form, we start with the following weighted residual form of (4.7).

$$\int\!\!\int_\Omega \left(\frac{\partial^2 \psi}{\partial x^2} + \frac{\partial^2 \psi}{\partial y^2} \right) \delta\psi \, d\Omega = 0 \, . \tag{4.9}$$

Integration by parts yields

$$\int\!\!\int_\Omega \left(\frac{\partial \psi}{\partial x} \frac{\partial \delta\psi}{\partial x} + \frac{\partial \psi}{\partial y} \frac{\partial \delta\psi}{\partial y} \right) d\Omega - \int_{\Gamma_s} \frac{\partial \psi}{\partial n} \delta\psi \, d\Gamma = 0 \, . \tag{4.10}$$

We insert the second boundary condition of (4.8) into this equation to obtain the weak form;

$$\int\!\!\int_\Omega \left(\frac{\partial \psi}{\partial x} \frac{\partial \delta\psi}{\partial x} + \frac{\partial \psi}{\partial y} \frac{\partial \delta\psi}{\partial y} \right) d\Omega = - \int_{\Gamma_s} V_s \, \delta\psi \, d\Gamma \, . \tag{4.11}$$

The domain is subdivided into triangular finite elements. The unknown ψ in an element e is linearly interpolated. The same interpolation functions are used for the weighting functions, namely,

$$\psi(x,y) \;=\; \sum_{\alpha=1}^{3} \phi_\alpha(x,y)\,\psi_\alpha \;,$$

(4.12)

$$\delta\psi(x,y) \;=\; \sum_{\alpha=1}^{3} \phi_\alpha(x,y)\,\delta\psi_\alpha \;,$$

where the interpolation functions ϕ_α are given by

$$\phi_\alpha(x,y) \;=\; \frac{1}{2\,\Delta^e}\,(\,a_\alpha + b_\alpha x + c_\alpha y\,)$$

(4.13)

with

$$a_\alpha \;=\; x_\beta y_\gamma - x_\gamma y_\beta\;, \quad b_\alpha \;=\; y_\beta - y_\gamma\;, \quad c_\alpha \;=\; x_\gamma - x_\beta\;,$$

(4.14)

$$\Delta^e \;=\; (\,b_\alpha c_\beta - b_\beta c_\alpha\,)/2\;,$$

in which the indices α,β,γ run in a cyclic order of 1, 2, 3.

The weak form (4.11) in each element e is thus discretised as

$$\sum_{\alpha=1}^{3} \iint_e \Big(\frac{\partial\psi}{\partial x}\frac{\partial\phi_\alpha}{\partial x} + \frac{\partial\psi}{\partial y}\frac{\partial\phi_\alpha}{\partial y}\Big)\,d\Omega\,\delta\psi_\alpha$$

$$= \;-\sum_{\alpha=1}^{3} \int_{\Gamma_s^e} V_s\,\phi_\alpha\,d\Gamma\,\delta\psi_\alpha \;,$$

(4.15)

where $\Gamma_s^e = \Gamma_s \cap \partial e$. From the arbitrariness of $\delta\psi_\alpha$, we obtain the following element equations.

$$\sum_{\beta=1}^{3} D_{\alpha\beta}^e\,\psi_\beta \;=\; F_\alpha^e \qquad (\alpha = 1,2,3)\;,$$

(4.16)

where the coefficients $D_{\alpha\beta}^e$ and the right-hand side terms F_α^e are given by

$$D_{\alpha\beta}^e \;=\; \iint_e \Big(\frac{\partial\phi_\alpha}{\partial x}\frac{\partial\phi_\beta}{\partial x} + \frac{\partial\phi_\alpha}{\partial y}\frac{\partial\phi_\beta}{\partial y}\Big)\,d\Omega\;,$$

(4.17)

$$F_\alpha^e \;=\; -\int_{\Gamma_s^e} V_s\,\phi_\alpha\,d\Gamma\;.$$

(4.18)

From (4.13), we can see that

$$[D^e] \;=\; \frac{1}{4\,\Delta^e}\begin{bmatrix} b_1^2 + c_1^2 & b_1 b_2 + c_1 c_2 & b_1 b_3 + c_1 c_3 \\ & b_2^2 + c_2^2 & b_2 b_3 + c_2 c_3 \\ \text{symmetric} & & b_3^2 + c_3^2 \end{bmatrix}\;,$$

(4.19)

and

$$\{F^e\} = -\frac{1}{2}V_s\,\ell \begin{bmatrix} 0 \\ 1 \\ 1 \end{bmatrix}. \qquad (4.20)$$

Here we assumed that V_s is constant on the boundary segment Γ_s^e, whose length is equal to ℓ in the situation illustrated in Figure 3.8.

After assembly of the element equations (4.16), we can obtain the total equation

$$[D]\{\psi\} = \{F\}. \qquad (4.21)$$

Once the boundary conditions of the Dirichlet type, given by (4.8), are inserted into (4.21), we can express the results in the form

$$[D_B]\{\psi\} = \{F_B\}. \qquad (4.22)$$

The nodal values of the streamfunction are obtained from the solution $\{\psi\}$ of this linear system of equations.

The velocity components u and v can be calculated by using (4.5), once the nodal values ψ_α are known, namely,

$$u = \sum_{\alpha=1}^{3} \frac{\partial \phi_\alpha}{\partial y}\psi_\alpha = \frac{1}{2\,\Delta^e}\sum_{\alpha=1}^{3} c_\alpha\,\psi_\alpha,$$

$$\hspace{8cm}(4.23)$$

$$v = -\sum_{\alpha=1}^{3} \frac{\partial \phi_\alpha}{\partial x}\psi_\alpha = -\frac{1}{2\,\Delta^e}\sum_{\alpha=1}^{3} b_\alpha\,\psi_\alpha.$$

4.3   Solution Procedure

A diagram of the solution procedure is depicted in Figure 4.1. Regardless of different formulations in terms of either the velocity potential or the streamfunction, the numerical solution proceeds as follows.

(a) *Data input: Topological data consisting of nodal coordinates and element connectivity, as well as the boundary conditions are read-in from an input data file.*

(b) *Total matrix: Calculate element matrices for all finite elements. Assemble all the element matrices to construct a total matrix.*

(c) *Boundary conditions: Insert the Dirichlet condition.*

(d) *Linear equation: Solve the linear system of equations to find nodal values of either velocity potential or streamfunction.*

(e) *Flow velocity: Calculate velocity components.*

(f) *Output the results: Write-out the calculated results on an output data file.*

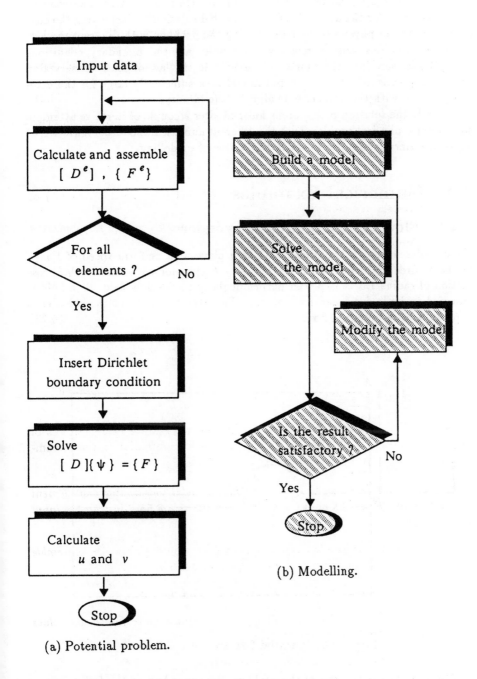

(a) Potential problem.

(b) Modelling.

Figure 4.1: Flow diagrams.

The first attempt at the solution procedure may not produce a satisfactory result. One may obtain an incorrect answer due to erroneous input data, inappropriate boundary conditions or due to unsuitable meshes. One may want to get a guarantee for the accuracy of his numerical solution by inspecting the conservation laws in physics or by recalculating the problem with different meshes.

Some questions arise herein. For example, are the boundary conditions plugged precisely into the numerical model? Is the law of mass conservation numerically satisfied? Is the computational area sufficiently large for the exterior problem with the domain extending to the infinity?

Usually the solution procedure is finished after a couple of iterative attempts for the same physical problem with the different numerical models. Figure 4.1(b) shows the iterative improvement in the course of modelling.

4.4 Numerical Examples

4.4.1 Flow around a circular cylinder

We consider a flow in two-dimensional x, y plane at a uniform speed of 1 m/s in the x direction, as shown in Figure 4.2. A cylinder with the circular cross section of radius 1.5 m is placed perpendicularly to the flow field. The fluid is assumed to be inviscid and irrotational. Then the flow around the cylinder is expressible as a potential flow.

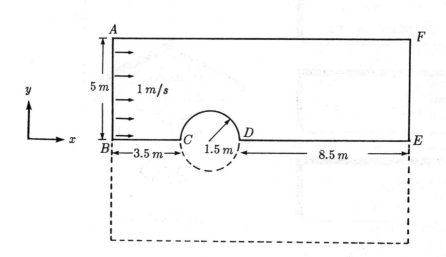

Figure 4.2: Potential flow around a cylinder.

Owing to the symmetry of the problem, an upper half of the full geometry will be considered. The part of the boundary AB represents the inlet, from which the fluid flows in at the velocity $u = 1\ m/s, v = 0\ m/s$, BC and DE are lines of symmetry. The semicircle CD is the slipping boundary, where the normal component of the flow velocity is equal to zero. EF represents the outlet taken sufficiently downstream, from which the fluid is assumed to flow out at the uniform velocity, or on which the tangential component of the velocity is assumed to be equal to zero. FA is a far field boundary, which is taken sufficiently remote from the obstacle so that the boundary forms a streamline.

We shall recast those boundary conditions in terms of the velocity potential Φ. From (4.1) we see that the location of the zero basis of Φ can be taken arbitrarily. We set $\Phi = 0$ along the inlet AB. Since the normal component of the velocity to BCDE and FA must be zero, we get $V_n = 0$ in (4.4). The distance between the inlet and the outlet is $15\ m$. From the first equality in the equations (4.1), the constant value of Φ on EF is given by

$$\Phi = \int_0^{15} u\,dx = 15\ m^2/s\ .$$

We shall recast the boundary conditions again in terms of the streamfunction ψ. From (4.5) we see that the location of the zero basis of ψ can be taken arbitrarily. We set $\psi = 0$ at the corner B. We notice that $u = 1\ m/s$, $v = 0\ m/s$ along AB. From the first equality in equations (4.5), the value of ψ on the line is given by

$$\psi(0, y) = \int_0^y u\,dy = y\ m^2/s\ .$$

At the corner A(0,5), we have $\psi = 5\ m^2/s$. Since BCDE forms a streamline, we get $\psi = 0$ there. AF forms another streamline with $\psi = 5$. The tangential component of the flow velocity is set to be equal to zero on EF. This implies that $V_s = 0\ m/s$ in (4.8).

To implement the finite element method, we subdivide the domain into 180 triangular finite elements with 115 nodes, as shown in Figure 4.3. Calculated results by using the velocity potential are presented in Figure 4.4. The inlet and outlet boundaries constitute the equipotential lines with the values $\Phi = 0$ and $\Phi = 15$, respectively. Other equipotential lines perpendicularly intersect the boundaries, along which the zero normal derivative of Φ is given.

Figure 4.5 shows calculated results by using the streamfunction. The lower and upper boundaries constitute streamlines with the values $\psi = 0$ and $\psi = 5$, respectively. Other streamlines perpendicularly intersect the outlet boundary, where the zero normal derivative of ψ is given. The calculated velocities presented in Figure 4.4(b) and 4.5(b) are in good agreement except those near the stagnation points, where the accuracy is lost due to the nature of the singularity. The set of equipotential lines intersects the set of streamlines perpendicularly, forming a curvilinear orthogonal net.

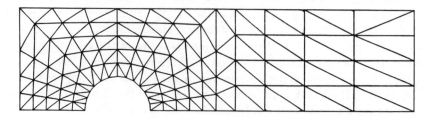

Figure 4.3: Finite element mesh.

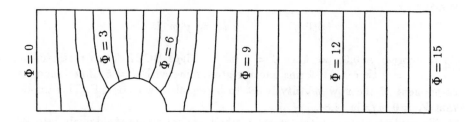

(a)   Equipotential lines.

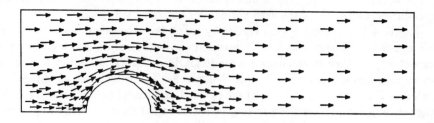

(b)   Flow velocity.

Figure 4.4: Calculated flow around a circular cylinder using the velocity potential.

$\psi = 5$

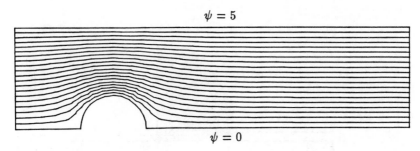

$\psi = 0$

(a) Streamlines.

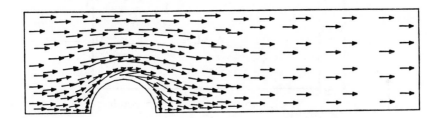

(b) Flow velocity.

Figure 4.5: Calculated flow around a cylinder using the streamfunction.

4.4.2   Groundwater flow

We consider the seepage of groundwater through the pores of soils in an aquifer. We assume that the compressibility of water is negligibly small. According to the D'Arcy law of seepage flow, the velocity components u and v are expressible in the form

$$u = -k\frac{\partial H}{\partial x}, \qquad v = -k\frac{\partial H}{\partial y}, \qquad (4.24)$$

where k is the hydraulic conductivity of the aquifer, and H is the piezometric head. We shall assume that k is a positive constant.

From the continuity equation (4.2), we obtain

$$k\left(\frac{\partial^2 H}{\partial x^2} + \frac{\partial^2 H}{\partial y^2}\right) = 0. \qquad (4.25)$$

This is the Laplace equation for unknown $H(x, y)$.

We shall give an example of the seepage flow under a dam, as shown in Figure 4.6. The embankment and the rock basement are assumed to be impermeable. The artificial boundaries AB and CD are introduced to confine the geometry in a bounded domain. The water depth of the right-hand side reservoir is equal to 35 m, while the depth of the left-hand side reservoir is 5 m.

Finite element mesh and associated boundary conditions are presented in

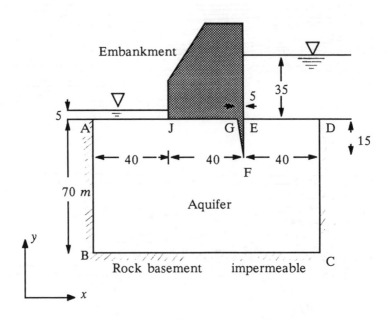

Figure 4.6: Seepage flow in a confined aquifer.

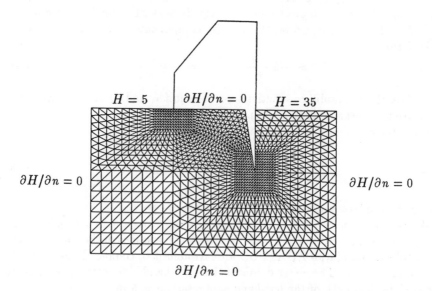

Figure 4.7: Mesh and boundary conditions for seepage flow.

Figure 4.7. The aquifer is subdivided into 2432 elements with 1281 nodes. The hydraulic conductivity is taken as $k = 10^{-6}\ m/s$.

Calculated results are shown in Figure 4.8. Isoclines perpendicularly intersect the impervious boundaries. As we will examine in the next section, the tip F of the pile and the base point J, indicated in Figure 4.6, are singular points in the sense that the unknown function H is continuous but the derivatives become infinite at these points. Mesh refinement in the vicinity of the singular points will improve the accuracy of the numerical solution to a greater degree.

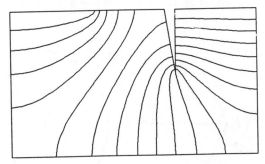

(a)   Piezometric head.

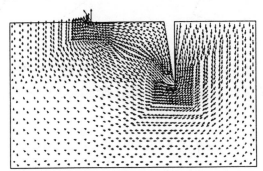

(b)   Seepage flow velocity.

Figure 4.8: Calculated seepage flow.

4.5   Singularities

We shall examine singularities of the solution $u(x, y)$ to the boundary value problem

$$- \Delta u = 0 \quad \text{in} \quad \Omega, \quad u = g(x, y) \quad \text{on} \quad \Gamma.$$

We assume that g can be extended to a smooth function u' on the whole domain Ω such that the extended u' takes the value g on the boundary Γ. Instead of considering u, we may consider the difference $u - u'$ as a new unknown. In this way, we can reformulate the problem as

$$-\Delta u \;=\; f(x,y) \quad \text{in} \quad \Omega, \tag{4.26}$$

$$u \;=\; 0 \quad \text{on} \quad \Gamma. \tag{4.27}$$

For the present purpose of discussion, it is sufficient that the geometry of Ω be confined to a sector with the angle $\alpha\pi$ ($0 < \alpha < 2$), as shown in Figure 4.9(a). Assume that f is analytic on the closure of the domain. From (4.26) we have

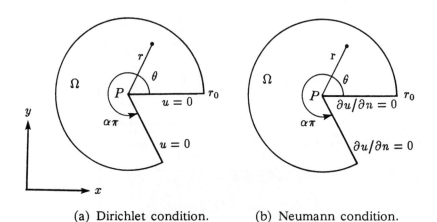

(a) Dirichlet condition.   (b) Neumann condition.

Figure 4.9: Pie-shaped domain with local polar coordinates.

$$\int_{\Omega} \nabla u \cdot \nabla v \, d\Omega \;=\; \int_{\Omega} f v \, d\Omega \tag{4.28}$$

for functions v, being arbitrary but $v = 0$ on Γ. With the polar coordinates as illustrated in Figure 4.9, the integral form (4.28) becomes

$$\int_{0}^{r_0} r \, dr \int_{0}^{\alpha\pi} \left(\frac{\partial u}{\partial r} \frac{\partial v}{\partial r} + \frac{1}{r^2} \frac{\partial u}{\partial \theta} \frac{\partial v}{\partial \theta} \right) d\theta \;=\; \int_{0}^{r_0} \int_{0}^{\alpha\pi} f v r \, dr \, d\theta. \tag{4.29}$$

Note that $u(r,0) = u(r,\alpha\pi) = 0$ from the boundary condition (4.27). We can express u by Fourier series of the form

$$u(r,\theta) \;=\; \sum_{j=1}^{\infty} u_j(r) \, \phi_j(\theta) \tag{4.30}$$

using the trigonometric functions

$$\phi_j(\theta) \;=\; \sqrt{\frac{2}{\alpha\pi}} \, \sin\left(j \frac{\theta}{\alpha}\right), \tag{4.31}$$

with the orthogonality property

$$\int_0^{\alpha\pi} \phi_j(\theta)\,\phi_k(\theta)\,d\theta \;=\; \delta_{jk} \;, \tag{4.32}$$

where δ_{jk} is the Kronecker delta. We notice the additional orthogonality;

$$\int_0^{\alpha\pi} \frac{d\phi_j}{d\theta}\frac{d\phi_k}{d\theta}\,d\theta \;=\; (\frac{j}{\alpha})^2\,\delta_{jk} \;. \tag{4.33}$$

We shall determine the Fourier coefficients $u_j(r)$. To this end, we set $v = w(r)\phi_k(\theta)$ for any function $w(r)$ with $w(r_0) = 0$ by taking (4.27) into account. From (4.29) we have

$$\int_0^{r_0} r\{\frac{du_k}{dr}\frac{dw}{dr} + \frac{1}{r^2}(\frac{k}{\alpha})^2 u_k\,w\}\,dr \;=\; \int_0^{r_0} f_k(r)\,w(r)\,r\,dr \;, \tag{4.34}$$

where we denote

$$f_k(r) \;=\; \int_0^{\alpha\pi} f(r,\theta)\,\phi_k(\theta)\,d\theta \;.$$

Integration of (4.34) by parts yields

$$\int_0^{r_0} \{\frac{d}{dr}(r\frac{du_k}{dr}) - \frac{1}{r}(\frac{k}{\alpha})^2 u_k + r\,f_k(r)\}\,w(r)\,dr \;=\; 0 \;. \tag{4.35}$$

Since w is chosen arbitrarily, it follows that

$$\frac{d}{dr}(r\frac{du_k}{dr}) - \frac{1}{r}(\frac{k}{\alpha})^2 u_k + r\,f_k(r) \;=\; 0 \;. \tag{4.36}$$

We know that the general solution of this second-order linear ordinary differential equation is given as the sum of a general solution of the homogeneous equation corresponding to (4.36) plus a particular solution of (4.36). To find the general solution of the homogeneous equation

$$\frac{d}{dr}(r\frac{du_k}{dr}) - \frac{1}{r}(\frac{k}{\alpha})^2 u_k \;=\; 0 \;. \tag{4.37}$$

We put

$$u_k \;=\; C\,r^{\rho_k}$$

with a constant C and the index ρ_k. By the substitutions of this expression into (4.37), we see that

$$\rho_k \;=\; \pm\frac{k}{\alpha} \;. \tag{4.38}$$

Hence, with arbitrary constants α_k and β_k, we arrive at

$$\alpha_k\,r^{k/\alpha} + \beta_k\,r^{-k/\alpha} \;, \tag{4.39}$$

which is the required solution. To find a particular solution of the non-homogeneous equation (4.36), we shall expand the analytic function $f_k(r)$ into the power series

$$f_k(r) \;=\; f_k^0 + f_k^1 r + f_k^2 r^2 + \cdots \tag{4.40}$$

with the coefficients $f_k^0, f_k^1, f_k^2, \cdots$. We seek out the solution of the form

$$u_k = C_k^0 + C_k^1 r + C_k^2 r^2 + \cdots . \qquad (4.41)$$

Substitutions of (4.40) and (4.41) into (4.36) result in

$$C_k^1 + 2^2 C_k^2 r + 3^2 C_k^3 r^2 + \cdots - (\frac{k}{\alpha})^2 (\frac{C_k^0}{r} + C_k^1 + C_k^2 r + C_k^3 r^2 + \cdots)$$

$$+ f_k^0 r + f_k^1 r^2 + f_k^2 r^3 + \cdots = 0 . \qquad (4.42)$$

Comparing the like powers of r, we can see that

$$C_k^0 = 0 \quad , \qquad\qquad C_k^1 = 0 \quad \text{or} \quad \frac{k}{\alpha} = 1 ,$$

$$C_k^2 = \frac{f_k^0}{(k/\alpha)^2 - 2^2} \quad , \qquad C_k^3 = \frac{f_k^1}{(k\alpha)^2 - 3^2}$$

$$\cdots\cdots \qquad\qquad (4.43)$$

$$C_k^n = \frac{f_k^{n-2}}{(k/\alpha)^2 - n^2} \quad , \qquad \cdots$$

When k/α happens to be equal to some integer $n (\geq 2)$, the coefficient C_k^n becomes indeterminate. In this case, we shall seek a solution of form other than (4.41). For this purpose, we put

$$u_k = C_k^0 + C_k^1 r + C_k^2 r^2 + \cdots$$
$$+ \ln r (D_k^0 + D_k^1 r + D_k^2 r^2 + \cdots) . \qquad (4.44)$$

By substituting (4.40) and (4.44) into (4.36), we have

$$C_k^1 + 2^2 C_k^2 r + 3^2 C_k^3 r^2 + \cdots - (\frac{k}{\alpha})^2 (\frac{C_k^0}{r} + C_k^1 + C_k^2 r + C_k^3 r^2 + \cdots)$$
$$+ 2(D_k^1 + 2D_k^2 r + 3D_k^3 r^2 + \cdots) + \ln r (D_k^1 + 2^2 D_k^2 r + 3^2 D_k^3 r^2 + \cdots)$$
$$- (\frac{k}{\alpha})^2 \ln r (\frac{D_k^0}{r} + D_k^1 + D_k^2 r + D_k^3 r^2 + \cdots)$$
$$+ f_k^0 r + f_k^1 r^2 + f_k^2 r^3 + \cdots = 0 . \qquad (4.45)$$

Comparing the like powers of r, we can see that

$$C_k^0 = 0 \quad , \qquad D_k^0 = 0 ,$$
$$C_k^1 = 0 \quad , \qquad D_k^1 = 0 , \qquad\qquad (4.46)$$
$$C_k^j = \begin{cases} \frac{f_k^{j-2}}{(k/\alpha)^2 - j^2} \\ 0 \end{cases} \quad , \quad D_k^j = \begin{cases} 0 & (j = n) \\ -\frac{f_k^{j-2}}{2j} & (j \neq n) \end{cases} .$$

Therefore, the general solution of the differential equation (4.36) is given by

$$u_k = \alpha_k r^{k/\alpha} + \beta_k r^{-k/\alpha}$$
$$+ \sum_{j=2}^{\infty} C_k^j r^j + \ln r \sum_{j=2}^{\infty} D_k^j r^j . \qquad (4.47)$$

Since u_k tends to zero as $r \to 0$, it is required that $\beta_k = 0$.

Thus, we know that the leading term of the solution has the form

$$u(r, \theta) = \alpha_1 \sqrt{\frac{2}{\alpha \pi}} r^{1/\alpha} \sin\left(\frac{\theta}{\alpha}\right) + \cdots \qquad (4.48)$$

with the non-zero constant α_1. It is easy to see that, when $\alpha > 1$, the first derivatives are not bounded as $r \to 0$. We say that the solution u has a *singularity of order* $r^{1/\alpha}$ at the corner P. Notice that the singularity comes out from the homogeneous equation (4.37). The inequality $\alpha > 1$ implies that P is a non-convex corner. When $\alpha = 3/2$, the corner is called a *re-entrant corner*. In the extreme case when $\alpha = 2$, the point P becomes a tip of the slit (a crack tip).

We now consider the singularity that occurs due to other boundary conditions than the Dirichlet condition given by (4.27). For this purpose, we shall consider the Neumann boundary value problem

$$-\Delta u = f(x, y) \quad \text{in} \quad \Omega, \qquad (4.49)$$

$$\frac{\partial u}{\partial n} = 0 \quad \text{on} \quad \Gamma, \qquad (4.50)$$

with the given analytic function f, as illustrated in Figure 4.9(b). To make the solution uniquely determined, we impose $u = 0$ at the corner P. The weak form (4.28) remains unchanged with arbitrary v. The equality (4.29) also remains true. From the boundary condition (4.50), we notice that

$$\frac{\partial u}{\partial \theta}(r, 0) = \frac{\partial u}{\partial \theta}(r, \alpha \pi) = 0, \quad \frac{\partial u}{\partial r}(r_0, \theta) = 0. \qquad (4.51)$$

Accordingly, the solution u of this Neumann problem is expressible in the Fourier series (4.30) using

$$\phi_j(\theta) = \sqrt{\frac{2}{\alpha \pi}} \cos\left(j \frac{\theta}{\alpha}\right), \qquad (4.52)$$

with the orthogonal properties

$$\int_0^{\alpha \pi} \phi_j(\theta) \phi_k(\theta) \, d\theta = \delta_{jk}, \qquad (4.53)$$

$$\int_0^{\alpha \pi} \frac{d\phi_j}{d\theta} \frac{d\phi_k}{d\theta} \, d\theta = \left(\frac{j}{\alpha}\right)^2 \delta_{jk}. \qquad (4.54)$$

By the similar arguments as was done for the foregoing Dirichlet problem, we can see that the leading term corresponding to the Neumann problem has the form

$$u(r, \theta) = \alpha_1 \sqrt{\frac{2}{\alpha \pi}} r^{1/\alpha} \cos\left(\frac{\theta}{\alpha}\right) + \cdots. \qquad (4.55)$$

It is easy to see again that, when $\alpha > 1$, the first derivatives are not bounded as $r \to 0$. In the example presented in Figure 4.6, it is found that the point F has the kind of singularity as (4.55). Correspondingly, the calculated fluxes are large in the finite elements near the non-convex corner F. Even if the mesh is much refined, the finite element solutions using the linear triangular finite elements will not converge to the exact solution near the singular point.

Exercises

4.1 Consider the Laplace equation (4.3) subject to the boundary conditions given by (4.4). Verify that

(i) the element equations are given by

$$\sum_{\beta=1}^{3} D_{\alpha\beta}^{e} \Phi_{\beta} = F_{\alpha}^{e} \qquad (\alpha = 1, 2, 3)$$

with

$$D_{\alpha\beta}^{e} = \frac{1}{4\Delta^{e}}(b_{\alpha} b_{\beta} + c_{\alpha} c_{\beta}),$$

$$F_{\alpha}^{e} = \int_{\Gamma_{n}^{e}} V_{n} \phi_{\alpha} \, d\Gamma,$$

(ii) the velocity components are given by

$$u = \sum_{\alpha=1}^{3} \frac{\partial \phi_{\alpha}}{\partial x} \Phi_{\alpha} = \frac{1}{2\Delta^{e}} \sum_{\alpha=1}^{3} b_{\alpha} \Phi_{\alpha},$$

$$v = \sum_{\alpha=1}^{3} \frac{\partial \phi_{\alpha}}{\partial y} \Phi_{\alpha} = \frac{1}{2\Delta^{e}} \sum_{\alpha=1}^{3} c_{\alpha} \Phi_{\alpha}.$$

4.2 Consider a two-dimensional incompressible flow around an airfoil, as shown in Figure 4.10. The flow is assumed to be irrotational. The governing equation in terms of the streamfunction is the Laplace equation (4.7). The

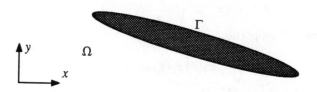

Figure 4.10: Exterior domain to an airfoil.

flow domain Ω is exterior to the boundary Γ of the airfoil cross section. To make the domain bounded, we consider an artificial boundary Γ_{∞}, which must be taken sufficiently remote from the airfoil in such a way that the flow is not disturbed on Γ_{∞} due to the presence of the airfoil.

The NACA four-digit airfoils can be obtained as follows: Let the four-digit code be described by $NACAd_1 d_2 d_3 d_4$. Put

$$m = 0.0d_1 = \frac{d_1}{100}, \qquad p = 0.d_2 = \frac{d_2}{10},$$

$$t = 0.d_3 d_4 = \frac{d_3}{10} + \frac{d_4}{100}.$$

Then the upper and lower surfaces of the airfoil with the unit length are given respectively by the curves

$$y(x) = y_c(x) \pm y_t(x), \qquad 0 \le x \le 1$$

where y_c and y_t express the center line and thickness of the airfoil, respectively, defined by

$$y_c = \begin{cases} \frac{m}{p^2}(2px - x^2), & 0 \le x \le p \\ \frac{m}{(1-p)^2}\{(1-2p) + 2px - x^2\}, & p \le x \le 1, \end{cases}$$

$$y_t = \frac{t}{0.20}(0.29690\sqrt{x} - 0.12600\,x - 0.35160\,x^2 + 0.28430\,x^3 - 0.10150\,x^4).$$

Consider the airfoil exposed to the flow at a uniform velocity of $u = U = 3.226\ m/s$, $v = 0\ m/s$. The condition on the far-field boundary is given by

$$\psi = -Uy \qquad \text{on} \qquad \Gamma_\infty.$$

The streamfunction on the surface of the airfoil must be constant, because the surface forms a streamline. However, the value of the streamfunction is not known *a priori*:

$$\psi = \gamma \qquad \text{on} \qquad \Gamma.$$

To determine the value γ, we shall paraphrase the problem into the following two separate problems:

$$\text{(I)} \qquad \Delta\psi_0 = 0 \quad \text{in} \quad \Omega,$$
$$\psi_0 = -Uy \quad \text{on} \quad \Gamma_\infty, \qquad \psi_0 = 0 \quad \text{on} \quad \Gamma;$$
$$\text{(II)} \qquad \Delta\psi_1 = 0 \quad \text{in} \quad \Omega,$$
$$\psi_1 = 0 \quad \text{on} \quad \Gamma_\infty, \qquad \psi_1 = 1 \quad \text{on} \quad \Gamma.$$

The solution is assumed to have the form

$$\psi = \psi_0 + \gamma\psi_1.$$

Chose a point Q near Γ_∞, where $\psi(Q)$ is known. Then the value of γ can be determined by the equation

$$\psi(Q) = \psi_0(Q) + \gamma\psi_1(Q).$$

Calculate streamlines for NACA3409 and compare your results with the followings: We have

$$-39.132992 = -39.221439 + \gamma \times 3.6338404,$$

$$\gamma = 2.433997.$$

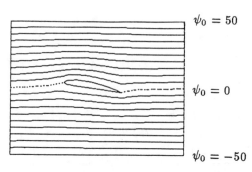

$$\psi_0 = 50$$

$$\psi_0 = 0$$

$$\psi_0 = -50$$

(a) Streamlines of ψ_0

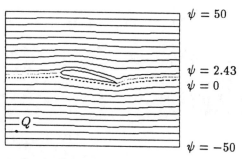

$$\psi = 50$$

$$\psi = 2.43$$
$$\psi = 0$$

$$\psi = -50$$

(b) Streamlines of ψ .

Figure 4.11: Calculated flows around the NACA3409 airfoil.

Figure 4.11 shows the calculated streamlines of ψ_0 and ψ. The Kutta's condition, which requires that the velocity at the trailing edge of the airfoil should be finite, is not considered here. Extend the above technique for the separation of problems to the solution of flows using the streamfunction in multiply connected domains.

4.3 Consider a seepage flow through an earth dam with a core, as shown in Figure 4.12. The governing equation is given by (4.25). The boundary BC forms a streamline, called the *phreatic surface* or *saturation surface*. We notice that the surface is a *free surface* in the sense that the location is not known *a priori*. The location must be determined as a part of the problem. The water seeps from the boundary CD. The baseline AE is assumed to be impervious. The associated boundary conditions are

$$H = H_a \quad \text{on} \quad AB \; ,$$

$$H = z \quad \text{and} \quad \frac{\partial H}{\partial n} = 0 \quad \text{on} \quad BC \; ,$$

$$\frac{\partial H}{\partial n} = \cos \alpha \quad \text{on} \quad CD \; ,$$

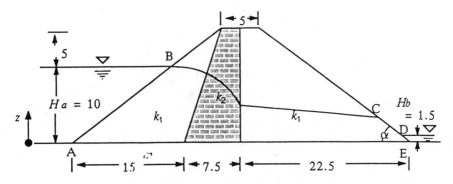

Figure 4.12: Seepage through an earth dam.

$$H = H_b \qquad \text{on} \qquad DE ,$$
$$\frac{\partial H}{\partial n} = 0 \qquad \text{on} \qquad EA .$$

Figure 4.13 shows the calculated equipotential lines in the case that $H_a = 10.0$ m, $H_b = 1.5$ m, $k_1 = 3.5 \times 10^{-6}$ m/s , and $k_2 = 2.1 \times 10^{-7}$ m/s. The phreatic surface is determined by an iterative procedure.

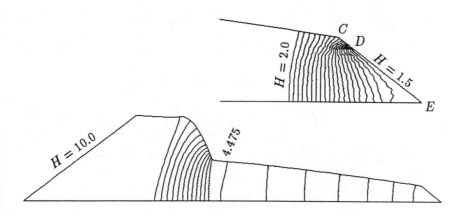

Figure 4.13: Calculated equipotential lines.

4.4 Derive an asymptotic expansion of the solution $H(x, y)$ of (4.25) using the polar coordinates near the point J in the Figure 4.6.

Hint : More generally, we shall consider the case illustrated in Figure 4.14. With the local polar coordinates centered at the point J, the boundary condition has the form
$$\frac{\partial H}{\partial \theta}(r, 0) = 0 , \qquad H(r, \alpha \pi) = 0 .$$

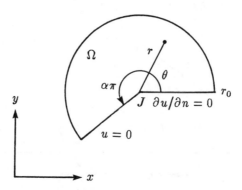

Figure 4.14: Pie-shaped domain with mixed boundary conditions.

In the neighbourhood of J, the solution can be expressed in the Fourier series

$$H(r, \theta) = \sum_{j=1}^{\infty} H_j(r)\, \phi_j(\theta) ,$$

using the trigonometric functions

$$\phi_j(\theta) = \sqrt{\frac{1}{\alpha\pi}}\, \cos\left(\frac{2j-1}{2\alpha}\theta\right) ,$$

with the orthogonal properties

$$\int_0^{\alpha\pi} \phi_j(\theta)\, \phi_k(\theta)\, d\theta = \delta_{jk} ,$$

$$\int_0^{\alpha\pi} \frac{d\phi_j}{d\theta}\frac{d\phi_k}{d\theta}\, d\theta = \frac{1}{2}\left(\frac{2j-1}{2\alpha}\right)^2 \delta_{jk} .$$

It turns out that the solution near the point J has the following asymptotic expansion.

$$H(r, \theta) = \alpha_1 \sqrt{\frac{1}{\alpha\pi}}\, r^{\frac{1}{2\alpha}}\, \cos\left(\frac{\theta}{2\alpha}\right) + \cdots .$$

We see that $\alpha = 1$ at the point J in Figure 4.6. The point J is a singular point due to an abrupt change in the type of boundary conditions. A steep gradient of the surface $H(x, y)$ at the singular point can be observed in Figure 4.15.

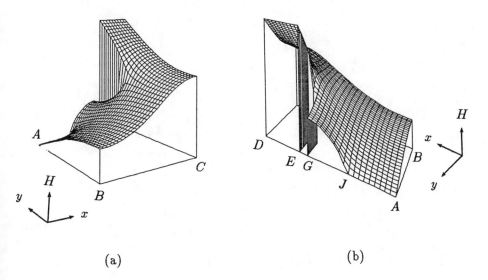

Figure 4.15: Perspective projection of the potential surface.

Chapter 5

TRANSIENT HEAT CONDUCTION

In this chapter we extend the finite element method to problems of transient heat conduction in a solid. The governing equation is the second-order parabolic equation. Finite element - Galerkin method is applied to the initial - boundary value problems in two dimensions. Discussions on stability and convergence of the numerical scheme are presented.

5.1 Basic Equations

We consider a thermally isotropic medium in the space with the rectangular coordinates (x, y, z). We assume now that the medium conducts heat in such a way that the temperature distribution is uniform with respect to the z direction. According to the *Fourier's law of heat conduction*, the components q_x, q_y $(J/m^2 \cdot s)$ of the *heat flux* in the x, y directions are proportional to the corresponding temperature gradients as follows.

$$q_x = -\kappa \frac{\partial T}{\partial x} , \qquad q_y = -\kappa \frac{\partial T}{\partial y} , \qquad (5.1)$$

where T is the temperature (K) and κ is the *heat conduction coefficient* $(J/m \cdot s \cdot K)$.

Let Ω be a cross section of the heat conductive medium, which is parallel to the xy-plane. The conservation of thermal energy is expressed in the form

$$\rho c \frac{\partial T}{\partial t} + \frac{\partial q_x}{\partial x} + \frac{\partial q_y}{\partial y} = 0 , \qquad (5.2)$$

where ρ is the density (kg/m^3), c is the *specific heat* $(J/kg \cdot K)$ of the medium, and t denotes the time variable. The product ρc is called the *heat capacity* $(J/m^3 \cdot K)$. Substituting (5.1) into (5.2), we obtain the *heat equation*

$$\rho c \frac{\partial T}{\partial t} = \nabla \cdot \kappa \nabla T . \qquad (5.3)$$

We assume that the temperature distribution in Ω at time t_0 is known as

$$T(x,y,t) = T^0(x,y) \quad \text{at} \quad t = t_0 . \tag{5.4}$$

This statement is called *initial condition*. The boundary of the domain Ω is denoted by Γ. The types of linear boundary conditions usually imposed are as follows.

(i) The temperature is specified on part of the boundary;

$$T = T_B \quad \text{on} \quad \Gamma_T . \tag{5.5}$$

(ii) The heat flux in the exterior normal direction is specified;

$$q_n = -\kappa \frac{\partial T}{\partial n} = q_B \quad \text{on} \quad \Gamma_q . \tag{5.6}$$

(iii) The heat radiation according to the Newton's law of cooling is described;

$$q_n = Q(T) = h(T - T_a) \quad \text{on} \quad \Gamma_h . \tag{5.7}$$

Here, we notice that T_B is the given temperature, q_B is the given boundary flux, h is the *thermal transmittivity* $(J/m^2 \cdot s \cdot K)$ of the boundary, and T_a is the ambient temperature. The problem consists of finding the unknown temperature $T(x,y,t)$ as a solution of the parabolic heat equation (5.3) subject to the initial boundary conditions (5.4)-(5.7).

5.2   Discretisation in Space and Time

Let δT be arbitrary weighting functions of the variables x and y, with $\delta T = 0$ on Γ_T, where the boundary temperature T is given by (5.5). To derive a weak form corresponding to the differential equation (5.3), we start with the weighted residual form

$$\int\int_\Omega (\rho c \frac{\partial T}{\partial t} - \nabla \cdot \kappa \nabla T) \delta T \, d\Omega = 0 . \tag{5.8}$$

We assume that the shape of Ω remains unchanged with time. Integration by parts yields

$$\int\int_\Omega \rho c \frac{\partial T}{\partial t} \delta T \, d\Omega - \int_\Gamma \kappa \frac{\partial T}{\partial n} \delta T \, d\Gamma$$
$$+ \int\int_\Omega \kappa \nabla T \cdot \nabla \delta T \, d\Omega = 0 , \tag{5.9}$$

where the vector operations indicate

$$\nabla T = (\frac{\partial T}{\partial x}, \frac{\partial T}{\partial y}) ,$$
$$\nabla T \cdot \nabla \delta T = \frac{\partial T}{\partial x} \frac{\partial \delta T}{\partial x} + \frac{\partial T}{\partial y} \frac{\partial \delta T}{\partial y} .$$

From the boundary conditions (5.6) and (5.7), we obtain the weak form

$$\iint_\Omega \rho c \frac{\partial T}{\partial t} \delta T \, d\Omega + \iint_\Omega \kappa \nabla T \cdot \nabla \delta T \, d\Omega$$

$$= - \int_{\Gamma_q} q_B \, \delta T \, d\Gamma - \int_{\Gamma_h} h \, (T - T_a) \delta T \, d\Gamma . \qquad (5.10)$$

We shall consider the discretisation of the weak form in the space. To this end, the domain Ω is subdivided into triangular finite elements. On each element e, the unknown temperature T and the weighting functions δT are approximated using the interpolation functions ϕ_α given by (4.13) as follows.

$$T(x, y, t) = \sum_{\alpha=1}^{3} \phi_\alpha(x, y) \, T_\alpha(t) ,$$

$$(5.11)$$

$$\delta T(x, y) = \sum_{\alpha=1}^{3} \phi_\alpha(x, y) \, \delta T_\alpha ,$$

where T_α are the time-dependent nodal temperatures, and δT_α are arbitrary nodal variations of T_α.

The weak form given by (5.10), as confined to each element e with (5.11), becomes

$$\sum_{\alpha=1}^{3} \iint_e \rho c \frac{\partial T}{\partial t} \phi_\alpha \, d\Omega \, \delta T_\alpha + \sum_{\alpha=1}^{3} \iint_e \kappa \nabla T \cdot \nabla \phi_\alpha \, d\Omega \, \delta T_\alpha$$

$$= - \sum_{\alpha=1}^{3} \int_{\Gamma_q^e} q_B \phi_\alpha \, d\Gamma \, \delta T_\alpha - \sum_{\alpha=1}^{3} \int_{\Gamma_h^e} h \, (T - T_a) \phi_\alpha \, d\Gamma \, \delta T_\alpha , \quad (5.12)$$

where $\Gamma_q^e = \Gamma_q \cap \partial e$ and $\Gamma_h^e = \Gamma_h \cap \partial e$. We assume that ρ, c, κ, h, T_a are constant elementwise. From the arbitrariness of δT_α, it results in

$$\rho c \sum_{\beta=1}^{3} M_{\alpha\beta}^e \frac{dT_\beta}{dt} + \kappa \sum_{\beta=1}^{3} D_{\alpha\beta}^e T_\beta$$

$$= - \Gamma_{q\alpha}^e - h \sum_{\beta=1}^{3} \gamma_{\alpha\beta}^e T_\beta + h \Gamma_{h\alpha}^e \qquad (\alpha = 1, 2, 3) \qquad (5.13)$$

with the coefficients

$$M_{\alpha\beta}^e = \iint_e \phi_\alpha \phi_\beta \, d\Omega = \frac{\Delta^e}{12} (1 + \delta_{\alpha\beta}) , \qquad (5.14)$$

$$D_{\alpha\beta}^e = \iint_e \left(\frac{\partial \phi_\alpha}{\partial x} \frac{\partial \phi_\beta}{\partial x} + \frac{\partial \phi_\alpha}{\partial y} \frac{\partial \phi_\beta}{\partial y} \right) d\Omega$$

$$= \frac{1}{4 \Delta^e} (b_\alpha b_\beta + c_\alpha c_\beta) , \qquad (5.15)$$

$$\Gamma^e_{q\alpha} \;=\; \int_{\Gamma^e_q} q_B\, \phi_\alpha\, d\Gamma \;=\; \frac{q_B}{2}\, |\,\Gamma^e_q\,| \begin{bmatrix} 0 \\ 1 \\ 1 \end{bmatrix}, \tag{5.16}$$

$$\gamma^e_{\alpha\beta} \;=\; \int_{\Gamma^e_h} \phi_\alpha\, \phi_\beta\, d\Gamma , \tag{5.17}$$

$$\Gamma^e_{h\alpha} \;=\; \int_{\Gamma^e_h} T_\alpha\, \phi_\alpha\, d\Gamma \;=\; \frac{T_\alpha}{2}\, |\,\Gamma^e_h\,| \begin{bmatrix} 0 \\ 1 \\ 1 \end{bmatrix}, \tag{5.18}$$

where $|\,\Gamma^e_q\,|$ and $|\,\Gamma^e_h\,|$ denote the lengths of Γ^e_q and Γ^e_h, respectively, for the similar situation presented in Figure 3.8.

The time derivative of the temperature can be approximated by using the finite difference as follows.

$$\dot{T}_\beta \;=\; \frac{dT_\beta}{dt} \approx \frac{T^{n+1}_\beta - T^n_\beta}{\Delta t}, \tag{5.19}$$

where T^n_β denotes the nodal temperature at the time level t_n, defined by $t_{n+1} = t_n + \Delta t\,(n = 0, 1, 2, \cdots)$ with the time increment Δt.

We consider an implicit scheme for (5.13) at the time level t_{n+1}, where the time derivative is replaced by (5.19). This results in

$$\rho c \sum_{\beta=1}^{3} M^e_{\alpha\beta} \frac{T^{n+1}_\beta - T^n_\beta}{\Delta t} + \kappa \sum_{\beta=1}^{e} D^e_{\alpha\beta}\, T^{n+1}_\beta$$

$$= -\Gamma^{e\,n+1}_{q\alpha} - h \sum_{\beta=1}^{e} \gamma^e_{\alpha\beta}\, T^{n+1}_\beta + h\,\Gamma^{e\,n+1}_{h\alpha} .$$

By the rearrangement, we have the following element equation:

$$\left(\frac{\rho c}{\Delta t} [M^e] + \kappa [D^e] + h[\gamma^e] \right) \{T^{n+1}\}$$

$$= \frac{\rho c}{\Delta t} [M^e] \{T^n\} - \{\Gamma^{e\,n+1}_q\} + h\{\Gamma^{e\,n+1}_h\} . \tag{5.20}$$

After assembly of (5.20) for all finite elements, we can obtain the total equation

$$\left(\frac{\rho c}{\Delta t} [M] + \kappa [D] + h[\gamma] \right) \{T^{n+1}\}$$

$$= \frac{\rho c}{\Delta t} [M] \{T^n\} - \{\Gamma^{n+1}_q\} + h\{\Gamma^{n+1}_h\} . \tag{5.21}$$

This is a linear recurrent system of equations for unknown $\{T^{n+1}\}$, provided $\{T^n\}$ is known. The initial nodal temperature $\{T^0\}$ is known from the initial condition given by (5.4). The sum of the coefficient matrices of $\{T^{n+1}\}$ is symmetric, positive definite, and it is often banded. After insertion of the Dirichlet boundary condition given by (5.5), the resulting equations can be solved step by step numerically for each increasing n.

The heat fluxes are obtained elementwise from (5.1), namely,

$$q_x = -\kappa \sum_{\alpha=1}^{3} \frac{\partial \phi_\alpha}{\partial x} T_\alpha = -\frac{\kappa}{2\,\Delta^e} \sum_{\alpha=1}^{3} b_\alpha T_\alpha ,$$

(5.22)

$$q_y = -\kappa \sum_{\alpha=1}^{3} \frac{\partial \phi_\alpha}{\partial y} T_\alpha = -\frac{\kappa}{2\,\Delta^e} \sum_{\alpha=1}^{3} c_\alpha T_\alpha .$$

5.3   Computational Procedure

As the Algol-like statements, the computation will proceed as follows.

Read *input data .*

Calculate coefficient matrices $[M]$, $[D]$, *and* $[\gamma]$.

For $n = 0, 1, 2, \cdots$, **until** *satisfied*, **do;**

> *Set* $\{T^n\}$ *onto the right-hand side of* (5.21).
> *Calculate the right-hand side using* $\{\Gamma_q^{n+1}\}$, $\{\Gamma_h^{n+1}\}$.
> *Insert the Dirichlet boundary condition.*
> *Solve the linear system of equations to find* $\{T^{n+1}\}$.
> *Calculate element fluxes.*

The algorithm is not complete until some stopping criteria are specified for the iteration counter n. When the time interval $[t_0, t_f]$ with the final time t_f, in which the temperature should be observed, is given, the counter n runs from zero up to $N = (t_f - t_0)/\Delta t$. Otherwise, the iteration may continue until the distribution of calculated temperatures attains a quasi-steady state. The commonly used criterion for this purpose is given by

$$\max_{\text{all nodes } \alpha} \left| \frac{T_\alpha^{n+1} - T_\alpha^n}{T_\alpha^n} \right| < \varepsilon$$

(5.23)

for some predetermined small $\varepsilon > 0$. The iteration may also be continued until the heat conduction fully develops from its transient state.

When will the conduction have fully developed? The answer to this question is attained by considering a non-dimensional form of the heat equation (5.3): For the sake of simplicity, we assume that the constant temperature T_B is prescribed on the whole boundary Γ and that the uniform initial temperature T^0 is given all over the domain Ω. Let L denote a representative length of the geometry. We introduce the non-dimensional variables x^*, y^* and T^* by the definition:

$$x^* = \frac{x}{L} , \quad y^* = \frac{y}{L} , \quad T^* = \frac{T - T^0}{T_B - T^0} .$$

(5.24)

We assume that κ is constant. Rewriting (5.3) by using these new variables, we obtain the corresponding non-dimensional form

$$\frac{\partial T^*}{\partial t^*} = (\frac{\partial^2}{\partial x^{*2}} + \frac{\partial^2}{\partial y^{*2}}) T^* \qquad (5.25)$$

with the non-dimensional time variable defined as

$$t^* = \frac{\kappa}{\rho c L^2} t = \frac{\lambda}{L^2} t . \qquad (5.26)$$

The non-dimensional time is called the Fourier time. In the book by H. Carslaw and J. Jaeger [1959], we know that figures of typical transient temperature distributions are often presented in the interval $0 < t^* < 1$. Any Fourier time greater than one is therefore an estimate, at which the temperature field attains almost quasi-steady state.

The numerical solution consists of finding the approximate temperatures T_α^n. The next question is how accurate are the T_α^n. A mathematical answer will be given in section 5.5. Here, we consider the problem of *consistency* and *stability* for the scheme (5.21). For this purpose, we shall take account of the following model equation in one dimension.

$$\frac{\partial u}{\partial t} = \lambda \frac{\partial^2 u}{\partial x^2} \quad , \quad -\infty < x < \infty , \ t > 0 . \qquad (5.27)$$

Let the nodes on the x axis be denoted by $x_j (j = 0, \pm 1, \pm 2, \cdots)$ with the increment Δx. The finite element discretisation using the linear interpolation functions given by (2.7) produces

$$\frac{1}{6} (\dot{u}_{j-1} + 4 \dot{u}_j + \dot{u}_{j+1}) = \lambda \frac{u_{j+1} - 2 u_j + u_{j-1}}{\Delta x^2} , \qquad (5.28)$$

where $u_j(t)$ denotes the approximate function to the exact $u(x_j, t)$. The left-hand side is a weighted average of the time-derivatives at three adjacent nodes, corresponding to Simpson's rule in the theory of numerical quadrature, while right-hand side is the central finite difference for the approximation of the second-order derivative.

Applying the approximation (5.19) to (5.28), we obtain the following implicit scheme.

$$\frac{1}{6} (\frac{u_{j-1}^{n+1} - u_{j-1}^n}{\Delta t} + 4 \frac{u_j^{n+1} - u_j^n}{\Delta t} + \frac{u_{j+1}^{n+1} - u_{j+1}^n}{\Delta t})$$
$$= \lambda \frac{u_{j+1}^{n+1} - 2 u_j^{n+1} + u_{j-1}^{n+1}}{\Delta x^2} , \qquad (5.29)$$

where u_j^n denotes the approximate value to the exact $u(x_j, t_n)$. This scheme is a linear system of difference equations. In order to assess the degree of resemblance between (5.27) and (5.29), we assume existence of a smooth function $U(x, t)$, satisfying the equation

$$\frac{1}{6} \left(\frac{U(x_{j-1}, t_{n+1}) - U(x_{j-1}, t_n)}{\Delta t} + 4\, \frac{U(x_j, t_{n+1}) - U(x_j, t_n)}{\Delta t} \right.$$
$$\left. + \frac{U(x_{j+1}, t_{n+1}) - U(x_{j+1}, t_n)}{\Delta t} \right)$$
$$= \lambda\, \frac{U(x_{j+1}, t_{n+1}) - 2\,U(x_j, t_{n+1}) + U(x_{j-1}, t_{n+1})}{\Delta x^2}. \tag{5.30}$$

Generally, the scheme is said to be *consistent* with the given differential equation if the corresponding differential equation, which any smooth solution $U(x, t)$ of the scheme satisfies, tends to the given differential equation as $\Delta x \to 0$ and $\Delta t \to 0$. We shall show that the scheme (5.29) is consistent with the differential equation (5.27). To this end, we expand U involved in (5.30) into the Taylor series around (x_j, t_n). After some algebra, we arrive at

$$\frac{1}{6} \left(\frac{U(x_{j-1}, t_{n+1}) - U(x_{j-1}, t_n)}{\Delta t} + 4\, \frac{U(x_j, t_{n+1}) - U(x_j, t_n)}{\Delta t} \right.$$
$$\left. + \frac{U(x_{j+1}, t_{n+1}) - U(x_{j+1}, t_n)}{\Delta t} \right)$$
$$= \frac{\partial U}{\partial t} + \frac{\Delta t}{2} \frac{\partial^2 U}{\partial t^2} + \frac{1}{6} \left(\Delta x^2 \frac{\partial^3 U}{\partial x^2 \partial t} + \Delta t^2 \frac{\partial^3 U}{\partial t^3} \right)$$
$$+ \frac{1}{24} \left(2\,\Delta x^2\, \Delta t\, \frac{\partial^4 U}{\partial x^2 \partial t^2} + \Delta t^3\, \frac{\partial^4 U}{\partial t^4} \right) + \cdots, \tag{5.31}$$

$$\frac{U(x_{j+1}, t_{n+1}) - 2\,U(x_j, t_{n+1}) + U(x_{j-1}, t_{n+1})}{\Delta x^2}$$
$$= \frac{\partial^2 U}{\partial x^2} + \Delta t\, \frac{\partial^3 U}{\partial x^2 \partial t} + \frac{1}{12} \left(\Delta x^2\, \frac{\partial^4 U}{\partial x^4} + 6\,\Delta t^2\, \frac{\partial^4 U}{\partial x^2 \partial t^2} \right) + \cdots, \tag{5.32}$$

where each term on the right-hand sides is to be evaluated at (x_j, t_n). Substituting these expansions into (5.30), we have

$$\frac{\partial U}{\partial t} = \lambda\, \frac{\partial^2 U}{\partial x^2} - \frac{\Delta t}{2} \frac{\partial^2 U}{\partial t^2} + \left(\lambda\, \Delta t - \frac{\Delta x^2}{6} \right) \frac{\partial^3 U}{\partial x^2 \partial t}$$
$$+ \frac{\lambda}{12}\, \Delta x^2\, \frac{\partial^4 U}{\partial x^4} + \cdots. \tag{5.33}$$

The terms abbreviated by $\cdots$ contain derivatives with the factors $\Delta x^\alpha \Delta t^\beta$ ($\alpha + \beta \geq 2$). Therefore, the scheme (5.29) proves to be consistent with (5.27), provided that all the derivatives of U are uniformly bounded.

The right-hand side of (5.33) involves not only spatial derivatives but also time derivatives. In order to transform the time derivatives to the corresponding spatial derivatives, we notice that

$$\frac{\partial}{\partial t} = \lambda\, \frac{\partial^2}{\partial x^2} - \frac{\Delta t}{2} \frac{\partial^2}{\partial t^2} + \cdots$$

$$= \lambda \frac{\partial^2}{\partial x^2} - \frac{\Delta t}{2}(\lambda \frac{\partial^2}{\partial x^2} - \frac{\Delta t}{2}\frac{\partial^2}{\partial t^2} + \cdots)^2 + \cdots$$

$$= \lambda \frac{\partial^2}{\partial x^2} - \frac{\lambda^2}{2}\Delta t \frac{\partial^4}{\partial x^4} + \cdots . \tag{5.34}$$

This implies that a first-order time derivative will yield infinitely many spatial derivatives of even-orders with a leading term of second-order. Thus, we can transform the right-hand side of (5.33) as follows.

$$\frac{\partial U}{\partial t} = \lambda \frac{\partial^2 u}{\partial x^2} - \frac{\Delta t}{2}\lambda^2 \frac{\partial^4 U}{\partial x^4} + (\lambda \Delta t - \frac{\Delta x^2}{6})\lambda \frac{\partial^4 U}{\partial x^4}$$

$$+ \frac{\lambda}{12}\Delta x^2 \frac{\partial^4 U}{\partial x^4} + \cdots$$

$$= \lambda \frac{\partial^2 U}{\partial x^2} + \frac{\Delta t}{2}\lambda^2 (1 - \frac{1}{6d})\frac{\partial^4 U}{\partial x^4} + \cdots , \tag{5.35}$$

where the non-dimensional number $d = \lambda \Delta t / \Delta x^2$ is called the *diffusion number* of the scheme. The right-hand side of (5.35) contains spatial derivatives of only even-orders.

The first term on the most right-hand side of (5.35) is the second-order derivative with the positive constant coefficient λ, which acts as conduction or diffusion. What is the function of the fourth-order derivative? To answer the question, we shall consider the equation

$$\frac{\partial U}{\partial t} = \lambda \frac{\partial^2 U}{\partial x^2} - \mu_2 \frac{\partial^4 U}{\partial x^4} , \quad -\infty < x < \infty , t > 0 \tag{5.36}$$

with some positive constant μ_2. Here in our case given by (5.35), $\mu_2 = \{(1/6d) - 1\}\Delta t \lambda^2/2$. We shall express the solution of (5.36) by the Fourier transform

$$U(x,t) = \int_{-\infty}^{+\infty} a(\sigma) e^{i(\omega t - \sigma x)} d\sigma , \tag{5.37}$$

where $a(\sigma)$ is the amplitude spectrum at the wave number $\sigma(1/m)$, ω is the angular frequency $(1/s)$, and i is the imaginary unit of complex numbers. Since $U(x,t)$ must be the solution of equation (5.36), the angular frequency is necessarily a function of the wave number: Substituting the Fourier transform into (5.36), by assuming that $a(\sigma)$ decays rapidly as $|\sigma| \to +\infty$, we can see that

$$i\omega = \lambda(-i\sigma)^2 - \mu_2(-i\sigma)^4 = -\lambda\sigma^2 - \mu_2\sigma^4 . \tag{5.38}$$

Therefore we have

$$U(x,t) = \int_{-\infty}^{+\infty} e^{(-\lambda\sigma^2 - \mu_2\sigma^4)t} a(\sigma) e^{-i\sigma x} d\sigma . \tag{5.39}$$

From this expression, we know that the amplitude at the wave number σ decays with the factor of $Exp[-(\lambda\sigma^2 + \mu_2\sigma^4)t]$, as time passes, and that the rate of decay is large at large wave numbers. In other words, the rate of decay becomes larger as the frequency becomes higher, provided that $\lambda + \mu_2 \sigma^2 > 0$. The positive μ_2 in (5.36) causes damping to each wave element. When μ_2 is negative, we see that $\lambda + \mu_2 \sigma^2 < 0$ for sufficiently large σ and that the solution $U(x,t)$ diverges exponentially for some non-zero initial values ultimately as $t \to +\infty$. This suggests that μ_2 should be made non-negative in the scheme: We have

$$d = \lambda \frac{\Delta t}{\Delta x^2} \leq \frac{1}{6} . \tag{5.40}$$

Although this condition can be satisfied for sufficiently small Δt, it is too strong for the practical implementation of the scheme. We remark that we have derived the condition by having discarded the terms of higher even-order derivatives than fourth-order in (5.35).

The scheme (5.29) is said to be *stable*, if the solution of (5.30) never diverges exponentially for any non-zero initial values. We notice that the condition (5.40) is neither necessary nor sufficient for stability. Let (5.35) be written as follows.

$$\frac{\partial U}{\partial t} = \lambda \frac{\partial^2 U}{\partial x^2} - \mu_2 \frac{\partial^4 U}{\partial x^4} + \mu_3 \frac{\partial^6 U}{\partial x^6} - \mu_4 \frac{\partial^8 U}{\partial x^8} \pm \cdots . \tag{5.41}$$

From similar arguments using the Fourier transform, as discussed above, we know that the scheme is stable if all the coefficients $\mu_k (k = 2, 3, 4, \cdots)$ are non-negative. More specifically, the scheme is stable if and only if $\lambda + \mu_2 \sigma^2 + \mu_3 \sigma^4 + \mu_4 \sigma^6 + \cdots \geq 0$ for all $\sigma(-\infty < \sigma < +\infty)$. However, this condition is difficult to check up.

Now, we shall derive the condition for stability in another way. To this end, we assume again that the solution of (5.30) is expressed in the form (5.37), where ω is a function of σ. In a time increment Δt, the solution becomes

$$U(x, t+\Delta t) = \int_{-\infty}^{+\infty} e^{i\omega\Delta t} a(\sigma) e^{i(\omega t - \sigma x)} d\sigma . \tag{5.42}$$

This implies that the amplitude $a(\sigma)$ is modulated by the factor of $Exp[i\omega\Delta t]$ after the short period Δt. The multiplier $Exp[i\omega\Delta t]$ is called the *amplification factor* of the scheme. The necessary and sufficient condition for the stability is

$$| e^{i\omega(\sigma)\Delta t} | \leq 1 \qquad \text{for } \forall \sigma . \tag{5.43}$$

If this inequality is satisfied for all $\Delta t > 0$, the scheme is said to be *unconditionally stable*. If the inequality is satisfied for some $\Delta t > 0$, then the scheme is said *conditionally stable*. The corresponding set of constraints on the magnitude of Δt is called the *stability condition*.

From scheme (5.29) and (5.30) we have

$$\frac{1}{6} \left(\frac{e^{i(\omega\Delta t + \sigma\Delta x)} - e^{i\sigma\Delta x}}{\Delta t} + 4 \frac{e^{i\omega\Delta t} - 1}{\Delta t} + \frac{e^{i(\omega\Delta t - \sigma\Delta x)} - e^{i\sigma\Delta x}}{\Delta t} \right)$$

$$= \lambda \frac{e^{i(\omega\Delta t - \sigma\Delta x)} - 2 e^{i\omega\Delta t} + e^{i(\omega\Delta t + \sigma\Delta x)}}{\Delta x^2} , \tag{5.44}$$

from which we can see that the amplification factor is given by

$$e^{i\omega(\sigma)\Delta t} = 1 \left/ \left(1 + 6d\frac{1 - \cos(\sigma\Delta x)}{2 + \cos(\sigma\Delta x)}\right)\right. . \qquad (5.45)$$

The denominator is always greater than unity on account of the inequalities $-1 \le \cos(\sigma\Delta x) \le 1$. Hence we know, as a conclusion, that the scheme (5.29) is unconditionally stable.

5.4   Numerical Examples

5.4.1   Heat conduction in a cooling cylinder

We consider the transient heat conduction in a thick circular cylinder with eight cooling ducts. Figure 5.1 shows a cross section of the cylinder. Inner and outer radii of the cylinder are 1 cm and 2 cm respectively. The radius of the cooling ducts is 0.2 cm.

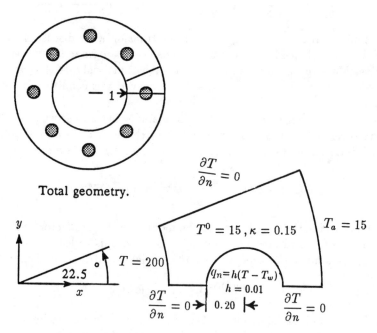

Total geometry.

Figure 5.1: 1/8 geometry of the cooling cylinder.

We assume that the physical constants of the material of the cylinder are $\rho c = 1\ Cal/cm^3 \cdot s \cdot {}^{\circ}C$ and $\kappa = 0.15\ Cal/cm \cdot s \cdot {}^{\circ}C$. A fluid at a constant temperature of 200 $^{\circ}C$ is flowing inside the cylinder, and the cylinder is exposed to surroundings at a temperature of $T_a = 15$ $^{\circ}C$. The cylinder is cooled though the surface of the cooling ducts by water at $T_w = 20$ $^{\circ}C$ according to Newton's law with $h = 0.01Cal/cm^2 \cdot s \cdot {}^{\circ}C$. At time $t = 0$, the temperature of the cylinder is $T^0 = 15\ °C$. The problem consists in tracing the transient distribution of temperatures inside the cylinder until the distribution attains the quasi-steady state. The time increment used is $\Delta t = 0.01\ s$.

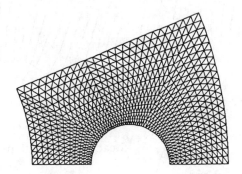

Figure 5.2: Finite element mesh.

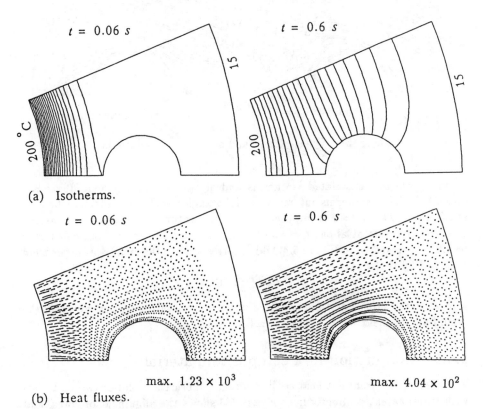

(a) Isotherms.

max. 1.23×10^3

max. 4.04×10^2

(b) Heat fluxes.

Figure 5.3: Calculated transient heat conduction.

Owing to the symmetry of the problem, only one-eighth of the total geometry may be considered in the computation. Figure 5.2 shows the triangular subdivision with 1344 elements and 741 nodes.

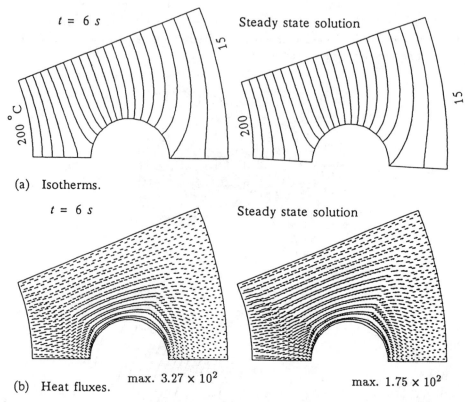

(a)   Isotherms.

(b)   Heat fluxes.   max. 3.27×10^2 max. 1.75×10^2

Figure 5.4: Quasi-steady and steady state solutions.

In Figure 5.3, calculated isotherms and the corresponding heat fluxes are presented. The isotherms intersect at right angles with the reflective boundaries, where $\partial T/\partial n = 0$ is imposed. The temperature distribution attains the quasi-steady state at about $t = 6\,s$. For the purpose of comparison, calculated results are presented in Figure 5.4, which correspond to the steady-state problem governed by the equation

$$\kappa \nabla^2 T = 0, \tag{5.46}$$

subject to the same set of boundary conditions given in Figure 5.1, by using the same finite element mesh as shown in Figure 5.2.

5.4.2   Conduction in a composite material

We consider the transient heat conduction in a fabricated solid of two materials with two different conductivities. Figure 5.5 shows the alignment of cores, and the domain of analysis and the boundary conditions of the problem as well: Owing to the cyclic geometry, a small portion of the composite material, which is marked with ABCD, is considered.   For the sake of simplicity, we assume in its non-dimensional form that $T = 1.0$ on the side AB and $T = 0.0$ on CD. The solid is assumed to be initially at a temperature of $T = 0.0$. The heat

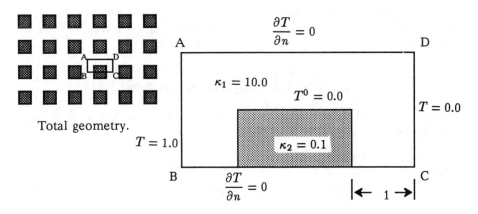

Figure 5.5: Core with the square cross section.

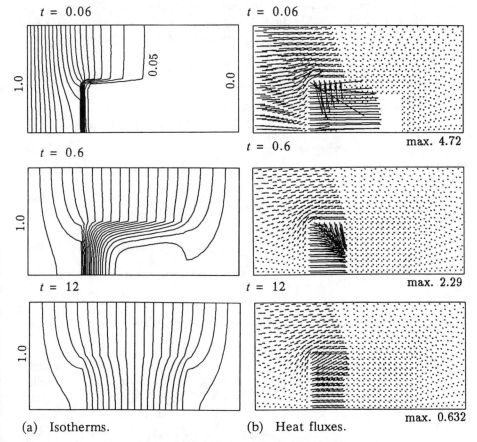

(a)   Isotherms.   (b)   Heat fluxes.

Figure 5.6: Evolution of calculated heat conduction.

conduction coefficients of the substrate and core are given as $\kappa_1 = 10.0$ and $\kappa_2 = 0.1$, respectively.

Calculated results are shown in Figure 5.6. The isotherms are twisted. The overall conductivity of the composite material is reduced due to the low conductivity of the core.

5.5   Convergence of Approximate Solutions

We shall present a brief discussion on the convergence of finite element solutions to the initial boundary value problem:

$$\frac{\partial u}{\partial t} - \Delta u = f \quad \text{in} \quad \Omega \times [0, \infty), \tag{5.47}$$

$$u = 0 \quad \text{on} \quad \Gamma \times [0, \infty), \tag{5.48}$$

$$u(x, +0) = u^0(x) \quad \text{in} \quad \Omega, \tag{5.49}$$

where Ω is a simply connected and bounded convex domain, enclosed by the smooth boundary Γ, in the plane with the rectangular coordinates $x = (x_1, x_2)$, and $f(x, t)$ is a given function. We assume for the sake of simplicity that the Cauchy data u^0 vanish on Γ. In this section, we shall summarize the basic mathematical results and the underlying arguments, which are mainly due to V. Thomée [1984].

We shall show the main result first: Let $u_h(x, t)$ denote a finite element approximate solution of the form

$$u_h(x, t) = \sum_{j=1}^{n} u_j(t) \phi_j(x) \tag{5.50}$$

with the finite element basis ϕ_j given by (3.12), and let $u_h^k(x)$ be the solution of the backward Euler-Galerkin equations

$$\left(\frac{u_h^{k+1} - u_h^k}{\Delta t}, \phi_i\right) + \left(\Delta u_h^{k+1}, \Delta \phi_i\right) = \left(f^{k+1}, \phi_i\right), \tag{5.51}$$

$$(i = 1, 2, \cdots, n)$$

with the $L^2(\Omega)$-scalar product $(\cdot, \cdot)$. The main result is stated in the next theorem.

Theorem 5.1 With the approximate initial values $u_h^0(x)$, the error in the L^2-norm $\|\cdot\|$ is bounded by

$$\|u_h^k - u(\cdot, t_k)\| \le \|u_h^0 - u^0\|$$

$$+ C h^2 \left\{ \|u^0\|_2 + \int_0^{t_k} \left\| \frac{\partial u}{\partial t}(\cdot, s) \right\|_2 ds \right\}$$

$$+ \Delta t \int_0^{t_k} \left\| \frac{\partial^2 u}{\partial t^2}(\cdot, s) \right\|_2 ds, \tag{5.52}$$

where $\|\cdot\|_2$ *denotes the norm equipped in the Sobolev space* $H_0^2(\Omega)$.

The error estimation given by the inequality (5.52) implies that, apart from the error committed in the discretisation of the initial values, the error in the approximate solutions $u_h^k(x)$, committed by the implicit scheme (5.51), is of order $O(h^2 + \Delta t)$, provided that the exact solution is sufficiently smooth. In the sequel we shall add some complements of the contents about the main theorem.

We subdivide Ω into triangular finite elements. Since Γ is a curved boundary in general, the union Ω_h of all the triangles does not coincide with Ω. However, we shall ignore such discrepancy. Let h denote the maximum diameter and let θ be the minimum angle among all the triangles. As the triangulation is made finer, we assume that θ is kept uniformly bounded below by some constant $\theta_0 > 0$ and that areas of the triangles are bounded below by ch^2 with a positive constant c which is independent of h.

Let n be the number of all the interior vertices P_j of Ω in the triangulation. We consider the finite element subspaces S_h of continuous functions on the closure of Ω:

$$S_h = \langle \phi_1, \phi_2, \cdots, \phi_n \rangle \subset C(\bar{\Omega})$$

with the piecewise linear pyramid functions ϕ_j defined by (3.12). By $L^2(\Omega)$ we shall denote the linear space of all the square Lebesque-summable functions on Ω, defined with the following scalar product and the corresponding norm:

$$(v, w) = \int_\Omega v(x) w(x) d\Omega, \qquad \|v\| = \sqrt{(v, v)}.$$

Moreover, by $H^m(\Omega)$ we shall denote the linear space of all the L^2-functions on Ω, whose derivatives of order up to m in the sense of distribution all belong to $L^2(\Omega)$. The space is defined with the norm denoted by

$$\|v\|_m = \sqrt{\sum_{0 \le \alpha_1 + \alpha_2 \le m} \left\| \frac{\partial^{\alpha_1 + \alpha_2} v}{\partial x_1^{\alpha_1} \partial x_2^{\alpha_2}} \right\|^2} \qquad (m = 1, 2, \cdots).$$

We shall introduce a subspace of $H_0^m(\Omega)$ by the definition:

$$H_0^m(\Omega) = \{ v \in H^m(\Omega) \, / \, v|_\Gamma = 0 \}.$$

For functions v in $H_0^2(\Omega)$, we consider the interpolation in S_h:

$$v^I(x) = \sum_{j=1}^n v(P_j) \phi_j(x).$$

It is known that

$$\|v^I - v\| \le C h^2 \|v\|_2, \tag{5.53}$$
$$\|\nabla v^I - \nabla v\| \le C h \|v\|_2, \tag{5.54}$$

with the generic constant $C > 0$. Therefore, it can be proved that, for $\forall v \in H_0^2(\Omega)$,

$$\inf_{\chi \in S_h} \{ \| v - \chi \| + h \| \nabla (v - \chi) \| \} \leq C h^2 \| v \|_2 . \tag{5.55}$$

We shall take ϕ from $H_0^2(\Omega)$. The exact solution $u(x,t)$ to (5.47) satisfies the weak form

$$(\frac{\partial u}{\partial t}, \phi) + (\nabla u, \nabla \phi) = (f, \phi) . \tag{5.56}$$

The semi-discrete solution $u_h(x,t)$ is implied by (5.50), satisfying

$$(\frac{\partial u_h}{\partial t}, \chi) + (\nabla u_h, \nabla \chi) = (f, \chi) \quad \text{for all} \quad \chi \in S_h , \tag{5.57}$$

together with the initial condition

$$u_h(x,0) = u_h^0(x) \quad \text{in} \quad S_h , \tag{5.58}$$

where u_h^0 is the projection of u^0 to S_h. The equations (5.57) are subjected to (5.58) to find unknown u_h which are equivalent to the ordinary differential equations

$$\sum_{j=1}^{n} (\phi_j, \phi_i) \dot{u}_j(t) + \sum_{j=1}^{n} (\nabla \phi_j, \nabla \phi_i) u_j(t) = (f, \phi_i) , \tag{5.59}$$

$$(i = 1, 2, \ldots, n)$$

together with the initial values

$$u_j(0) = u_h^0(P_j) , \quad (j = 1, 2, \ldots, n) \tag{5.60}$$

to find unknown functions $u_j(t)$. Since this section is devoted to a brief description of the convergence properties, we will state the next theorem without proof.

Theorem 5.2 *The error of the semi-discrete solution in L^2-norm is bounded by*

$$\| u_h(\cdot, t) - u(\cdot, t) \| \leq \| u_h^0 - u^0 \|$$

$$+ C h^2 \{ \| u^0 \|_2 + \int_0^t \| \frac{\partial u}{\partial t}(\cdot, s) \|_2 \, ds \} . \tag{5.61}$$

We shall discretise $u_j(t)$ with respect to the time variable by introducing the breaks $t_{k+1} = t_k + \Delta t$ $(k = 0, 1, 2, \ldots; \ t_0 = 0)$ with the time increment Δt. The time derivatives involved in (5.59) are approximated by the backward finite differences. This yields the following implicit scheme.

$$\sum_{j=1}^{n} (\phi_j, \phi_i) \frac{u_j^{k+1} - u_j^k}{\Delta t} + \sum_{j=1}^{n} (\nabla \phi_j, \nabla \phi_i) u_j^{k+1} = (f^{k+1}, \phi_i) , \tag{5.62}$$

$$\text{with} \quad u_j^0 = u_j(0) , \quad (i = 1, 2, \ldots, n) .$$

The Theorem 5.1 states that the sequence of the approximate solutions

$$u_h^k(x) \; = \; \sum_{j=1}^{n} u_j^k \, \phi_j(x) \, , \qquad (k = 1, 2, \ldots)$$

satisfies (5.51) and the error is estimated by (5.52).

Exercises

5.1 Solve the problem presented in Subsection 5.4.2 as $k_2 \to 0$.

5.2 Show that the point E in Figure 5.7 is a singular point.

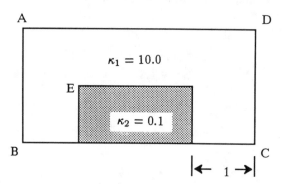

Figure 5.7: Problem with internal singular point.

5.3 We consider the vertical flow of a viscous fluid under a given pressure difference in three-dimensional space with the rectangular coordinates (x, y, z).
Suppose that the streamlines are parallel to the z-axis. The velocity vector has the three components $(u, v, w) = (0, 0, w)$. Then the continuity
equation reduces to

$$\frac{\partial w}{\partial z} = 0 .$$

This implies that $w = w(x, y, t)$, being a function of only x, y and t. Let
$\alpha(> 0)$ be the pressure gradient. Namely,

$$\frac{\partial p}{\partial z} = -\alpha .$$

From the Navier-Stokes equations we have

$$\frac{\partial p}{\partial x} = 0 \quad , \quad \frac{\partial p}{\partial y} = 0 ,$$

$$\frac{\partial w}{\partial t} = \nu \left(\frac{\partial^2}{\partial x^2} + \frac{\partial^2}{\partial y^2} \right) w + \frac{\alpha}{\rho} ,$$

where ν and ρ are the kinematic viscosity and density of the fluid, respectively. We know that $p = p(z, t)$. Using $\nu = 1.16 \times 10^{-7} \ m^2/s$, $\rho =
1.35 \times 10^4 \ kg/m^3$ for mercury at 20 °C, calculate the velocity w of the uni-
directional flow between a bundle of fission tubes, as shown in Figure 5.8,
subject to the pressure gradient $\alpha = 2.02 \times 10^3 \ Pa/m \ (= 0.02 \ atm/m)$.

Hint : Figure 5.9 shows a calculated result.

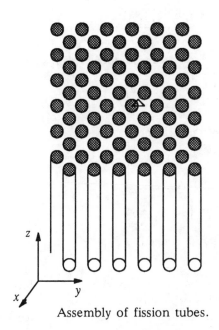

Assembly of fission tubes.

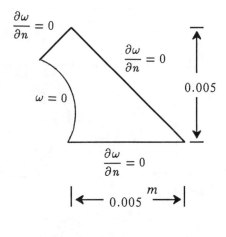

Problem statement.

Figure 5.8: Uni-directional viscous fluid flow.

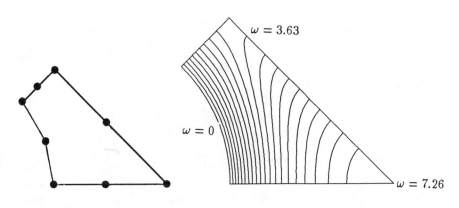

Block mesh.

Figure 5.9: Calculated velocity contours.

5.4 The left-hand side of (5.28) is a weighted average in the time derivatives. We scrape together the weights to the central node to obtain the following equation with the *lumped mass*.

$$\dot{u}_j = \lambda \frac{u_{j+1} - 2u_j + u_{j-1}}{\Delta x^2} .$$

Applying the Euler's approximation, we can get the lumped explicit scheme:

$$\frac{u_j^{n+1} - u_j^n}{\Delta t} = \lambda \frac{u_{j+1}^n - 2u_j^n + u_{j-1}^n}{\Delta x^2} .$$

Examine the accuracy and stability of this scheme.

Chapter 6

INCOMPRESSIBLE VISCOUS FLOW

In this chapter, we shall present the finite element method applied to problems of two-dimensional laminar flow of incompressible viscous fluid. The governing equations of the fluid motion are expressed in terms of the streamfunction and vorticity. For the analysis of time-dependent flows, we shall consider a semi-implicit time-marching scheme. The time integration is continued until the flow fully develops. Specific boundary conditions to vorticity are discussed. We shall show some examples of flows over a back step, around a cylinder, and in a square cavity.

6.1 Governing Equations

We shall consider an unsteady flow in the two-dimensional rectangular coordinates x, y. The velocity components in the x, y directions are denoted by u, v respectively. The Navier-Stokes equations can be written in the form

$$\frac{\partial \omega}{\partial t} + u \frac{\partial \omega}{\partial x} + v \frac{\partial \omega}{\partial y} = \nu \Delta^2 \omega , \qquad (6.1)$$

where ν is the kinematic viscosity, and ω is the vorticity defined by

$$\omega = \frac{\partial v}{\partial x} - \frac{\partial u}{\partial y} . \qquad (6.2)$$

When there are no sinks and sources inside the flow domain Ω, the equation of continuity takes the form:

$$\frac{\partial u}{\partial x} + \frac{\partial v}{\partial y} = 0 . \qquad (6.3)$$

Using the streamfunction ψ, this equation is automatically satisfied when the next relations are substituted:

$$u = \frac{\partial \psi}{\partial y}, \qquad v = -\frac{\partial \psi}{\partial x}. \tag{6.4}$$

From (6.2) and (6.4), we have

$$\frac{\partial^2 \psi}{\partial x^2} + \frac{\partial^2 \psi}{\partial y^2} = -\omega. \tag{6.5}$$

Suppose that the vorticity on the right-hand side is known. Then, this equation can be regarded as *Poisson equation* for the unknown streamfunction. Suppose moreover that the flow is irrotational, *i.e.*, $\omega = 0$. Then, the equation is reduced to the Laplace equation:

$$\frac{\partial^2 \psi}{\partial x^2} + \frac{\partial^2 \psi}{\partial y^2} = 0. \tag{6.6}$$

Replacing the velocity components u and v, which appeared in (6.1), with the corresponding relations given by (6.4), we have

$$\frac{\partial \omega}{\partial t} + \frac{\partial \psi}{\partial y}\frac{\partial \omega}{\partial x} - \frac{\partial \psi}{\partial x}\frac{\partial \omega}{\partial y} = \nu \Delta^2 \omega. \tag{6.7}$$

We shall consider typical boundary conditions, as illustrated in Figure 6.1. The conditions on the streamfunction are; $\psi = \psi_B$ which is specified on the boundary Γ_ψ, and $\partial\psi/\partial n = -V_s$ with the tangential flow velocity V_s specified on the rest of the boundary Γ_s. Here n denotes the outward unit normal to the boundary.

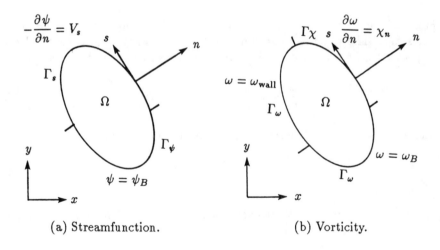

(a) Streamfunction.   (b) Vorticity.

Figure 6.1: Boundary conditions.

Boundary conditions on the vorticity are: $\omega = \omega_B$ which is specified on the boundary Γ_ω, and $\partial\omega/\partial n = \chi_n$ with the value of the normal derivative specified on the other part of the boundary Γ_χ. In addition, some other conditions on the rigid boundary Γ_w will be explained in Section 6.3.

6.2   Finite Element Discretisation

We shall apply the Galerkin method to spatial discretisation of the streamfunction and vorticity. To save space, we shall denote double integrals merely by using one integral sign hereafter. Let $\delta\psi$ and $\delta\omega$ be weighting functions, which are arbitrary but $\delta\psi = 0$ on Γ_ψ, $\delta\omega = 0$ on $\Gamma_\omega \cup \Gamma_w$. The weighted residual forms of (6.5) and (6.7) can be written as follows.

$$\int_\Omega \delta\psi \, \nabla^2\psi \, d\Omega \;+\; \int_\Omega \delta\psi \, \omega \, d\Omega \;=\; 0 \,, \tag{6.8}$$

$$\int_\Omega \delta\omega \, \frac{\partial\omega}{\partial t} \, d\Omega \;+\; \int_\Omega \delta\omega \left(\frac{\partial\psi}{\partial y}\frac{\partial\omega}{\partial x} - \frac{\partial\psi}{\partial x}\frac{\partial\omega}{\partial y} \right) d\Omega \tag{6.9}$$

$$-\int_\Omega \delta\omega \, \nu \, \nabla^2\omega \, d\Omega \;=\; 0 \,.$$

Integration by parts yields

$$\int_\Omega \nabla\delta\psi \cdot \nabla\psi \, d\Omega - \int_\Omega \delta\psi \, \omega \, d\Omega - \int_{\Gamma_s} \delta\psi \, \frac{\partial\psi}{\partial n} \, d\Gamma \;=\; 0 \,, \tag{6.10}$$

$$\int_\Omega \delta\omega \, \frac{\partial\omega}{\partial t} \, d\Omega + \int_\Omega \delta\omega \left(\frac{\partial\psi}{\partial y}\frac{\partial\omega}{\partial x} - \frac{\partial\psi}{\partial x}\frac{\partial\omega}{\partial y} \right) d\Omega \tag{6.11}$$

$$+ \int_\Omega \nu \nabla\delta\omega \cdot \nabla\omega \, d\Omega - \int_{\Gamma_\chi} \nu \, \delta\omega \, \frac{\partial\omega}{\partial n} \, d\Gamma \;=\; 0 \,.$$

The unknown ψ, ω and the corresponding first variations $\delta\psi$, $\delta\omega$ are approximated by using the linear shape functions ϕ_α on each triangular finite element e as follows:

$$\psi \;=\; \sum_{\alpha=1}^{3} \phi_\alpha \psi_\alpha \,, \qquad \delta\psi \;=\; \sum_\alpha \phi_\alpha \, \delta\psi_\alpha \,,$$

$$\tag{6.12}$$

$$\omega \;=\; \sum_{\alpha=1}^{3} \phi_\alpha \omega_\alpha \,, \qquad \delta\omega \;=\; \sum_\alpha \phi_\alpha \, \delta\omega_\alpha \,.$$

Here, ϕ_α and ω_α are nodal values of the streamfunction and vorticity, respectively. The shape function is given in the form:

$$\phi_\alpha \;=\; \frac{1}{2\,\Delta^e} \left(a_\alpha + b_\alpha x + c_\alpha y \right)$$

with the area Δ^e of the triangular finite element, and with the coefficients;

$$a_1 \;=\; x_2 y_3 - x_3 y_2 \,, \qquad b_1 \;=\; y_2 - y_3 \,, \qquad c_1 \;=\; x_3 - x_2 \,,$$

$$a_2 = x_3 y_1 - x_1 y_3, \quad b_2 = y_3 - y_1, \quad c_2 = x_1 - x_3,$$
$$a_3 = x_1 y_2 - x_2 y_1, \quad b_3 = y_1 - y_2, \quad c_3 = x_2 - x_1,$$

for a typical triangular element, as illustrated in Figure 6.2.

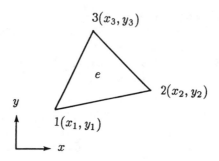

Figure 6.2: Triangular finite element.

We substitute (6.12) into the weak forms (6.10) and (6.11). From the arbitrariness in the nodal variations $\delta\psi_\alpha$, $\delta\omega_\alpha$, we can obtain the following *element equations*.

$$\sum_{\beta=1}^{3} D_{\alpha\beta}^e \, \psi_\beta - \sum_{\beta=1}^{3} M_{\alpha\beta}^e \, \omega_\beta - \Gamma_{s\alpha}^e = 0 , \qquad (6.13)$$

$$\sum_{\beta=1}^{3} M_{\alpha\beta}^e \, \dot\omega_\beta + \sum_{\beta=1}^{3} A_{\alpha\beta}^e \, \omega_\beta + \nu \sum_{\beta=1}^{3} D_{\alpha\beta}^e \, \omega_\beta - \Gamma_{\chi\alpha}^e = 0 , \qquad (6.14)$$

for $\alpha = 1, 2, 3$. Here we can see that

$$M_{\alpha\beta}^e = \int_e \phi_\alpha \, \phi_\beta \, d\Omega = \frac{\Delta^e}{12} (1 + \delta_{\alpha\beta}) , \qquad (6.15)$$

$$D_{\alpha\beta}^e = \int_e \left(\frac{\partial\phi_\alpha}{\partial x} \frac{\partial\phi_\beta}{\partial x} + \frac{\partial\phi_\alpha}{\partial y} \frac{\partial\phi_\beta}{\partial y} \right) d\Omega$$

$$= \frac{1}{4\,\Delta^e} (b_\alpha b_\beta + c_\alpha c_\beta) , \qquad (6.16)$$

$$A_{\alpha\beta}^e = \int_e \phi_\alpha \left(\sum_{\gamma=1}^{3} \frac{\partial\phi_\gamma}{\partial y} \psi_\gamma \frac{\partial\phi_\beta}{\partial x} - \sum_{\gamma=1}^{3} \frac{\partial\phi_\gamma}{\partial x} \psi_\gamma \frac{\partial\phi_\beta}{\partial y} \right) d\Omega , \qquad (6.17)$$

$$\Gamma_{s\alpha}^e = \int_{\Gamma_s^e} \phi_\alpha \frac{\partial\psi}{\partial n} \, d\Gamma ,$$

$$\Gamma_{\chi\alpha}^e = \int_{\Gamma_\chi^e} \nu \, \phi_\alpha \frac{\partial\omega}{\partial n} \, d\Gamma ,$$

where $\Gamma_s^e = \Gamma_s \cap \partial e$, $\Gamma_\chi^e = \Gamma_\chi \cap \partial e$, and $\delta_{\alpha\beta}$ is the Kronecker's delta. The coefficients $A^e_{\alpha\beta}$ are given explicitly in the matrix form:

$$[A^e] = \frac{1}{6} \begin{bmatrix} \psi_2 - \psi_3 & \psi_3 - \psi_1 & \psi_1 - \psi_2 \\ \psi_2 - \psi_3 & \psi_3 - \psi_1 & \psi_1 - \psi_2 \\ \psi_2 - \psi_3 & \psi_3 - \psi_1 & \psi_1 - \psi_2 \end{bmatrix} . \tag{6.18}$$

We shall prove (6.18): From (6.17), we see that

$$\sum_{\beta=1}^{3} A^e_{\alpha\beta} \omega_\beta = \sum_{\beta=1}^{3} \int_e \phi_\alpha \left(\sum_{\gamma=1}^{3} \frac{\partial \phi_\gamma}{\partial y} \psi_\gamma \frac{\partial \phi_\beta}{\partial x} - \sum_{\gamma=1}^{3} \frac{\partial \phi_\gamma}{\partial x} \psi_\gamma \frac{\partial \phi_\beta}{\partial y} \right) d\Omega \, \omega_\beta$$

$$= \sum_{\beta=1}^{3} \int_e \phi_\alpha \left(\sum_{\gamma=1}^{3} \frac{c_\gamma}{2\,\Delta^e} \psi_\gamma \frac{b_\beta}{2\,\Delta^e} - \sum_{\gamma=1}^{3} \frac{b_\gamma}{2\,\Delta^e} \psi_\gamma \frac{c_\beta}{2\,\Delta^e} \right) d\Omega \, \omega_\beta$$

$$= \sum_{\beta=1}^{3} \int_e \phi_\alpha \, d\Omega \left(\frac{1}{2\,\Delta^e} \right)^2 \sum_{\gamma=1}^{3} (c_\gamma b_\beta - b_\gamma c_\beta) \psi_\gamma \, \omega_\beta .$$

Using the following relations;

$$\int_e \phi_\alpha \, d\Omega = \frac{\Delta^e}{3} ,$$

and

$$2\,\Delta^e = c_1 b_3 - c_3 b_1 = c_2 b_1 - c_1 b_2 = c_3 b_2 - c_2 b_3 ,$$

we can obtain

$$\sum_{\beta=1}^{3} A^e_{\alpha\beta} \omega_\beta = \frac{1}{12\Delta^e} \Big\{ \sum_{\gamma=1}^{3} (c_\gamma b_1 - b_\gamma c_1) \psi_\gamma \omega_1$$

$$+ \sum_{\gamma=1}^{3} (c_\gamma b_2 - b_\gamma c_2) \psi_\gamma \omega_2 + \sum_{\gamma=1}^{3} (c_\gamma b_3 - b_\gamma c_3) \psi_\gamma \omega_3 \Big\}$$

$$= \frac{1}{12\Delta^e} \Big\{ 2\Delta^e (\psi_2 - \psi_3) \omega_1 + 2\Delta^e (\psi_3 - \psi_1) \omega_2$$

$$+ 2\Delta^e (\psi_1 - \psi_2) \omega_3 \Big\}$$

$$= \frac{1}{6} \big\{ (\psi_2 - \psi_3) \omega_1 + (\psi_3 - \psi_1) \omega_2 + (\psi_1 - \psi_2) \omega_3 \big\} .$$

Notice that the last result does not depend on the index α. This completes the proof.

We assemble (6.13) and (6.14) with respect to all the finite elements to construct equations on the whole domain. The resulting equations can be written in the matrix forms:

$$[D]\{\psi\} - [M]\{\omega\} - \{\Gamma_s\} = \{0\} , \tag{6.19}$$

$$[M]\{\dot{\omega}\} + [A(\psi)]\{\omega\} + \nu[D]\{\omega\} - \{\Gamma_\chi\} = \{0\} . \tag{6.20}$$

Let t_n be the time steps, defined by $t_{n+1} = t_n + \Delta t$ ($n = 0, 1, 2, \ldots$) with the time increment Δt. We denote the values of the streamfunction and vorticity at the time step t_n by $\{\psi^n\}$ and $\{\omega^n\}$, respectively. The time derivative involved in (6.20) is approximated by the finite difference:

$$\{\dot{\omega}\} \approx \frac{1}{\Delta t}(\{\omega^{n+1}\} - \{\omega^n\}). \qquad (6.21)$$

Then we can obtain the following one-step semi-implicit scheme.

$$[D]\{\psi^{n+1}\} = [M]\{\omega^n\} + \{\Gamma_s^n\}, \qquad (6.22)$$

$$\frac{1}{\Delta t}[M]\{\omega^{n+1}\} + \nu[D]\{\omega^{n+1}\}$$
$$= \frac{1}{\Delta t}[M]\{\omega^n\} - [A(\psi^{n+1})]\{\omega^n\} + \{\Gamma_\chi^{n+1}\}. \qquad (6.23)$$

The coefficient matrices on the left-hand sides of these equations are symmetric and are often banded. This property enables us to apply the symmetric band-matrix solver for the numerical solution of the linear system of equations.

Components of the flow velocity can be calculated inside each of the triangular finite elements, using (6.4);

$$u = \frac{1}{2\Delta^e} \sum_{\alpha=1}^{3} c_\alpha \psi_\alpha, \qquad (6.24)$$

$$v = -\frac{1}{2\Delta^e} \sum_{\alpha=1}^{3} b_\alpha \psi_\alpha. \qquad (6.25)$$

6.3   Boundary Conditions of the Vorticity

Boundary values of the vorticity on the wall and on the free surface are not known *a priori*. Those values must be calculated as a part of the problem. The calculated values denoted by ω_{wall} are used in the same way as ψ_B in the essential boundary condition.

We shall consider the vicinity of a boundary point B and an internal point N adjacent to it at distance Δy in the fluid, as illustrated in Figure 6.3. Suppose that the streamfunction ψ is sufficiently smooth. Then it can be expanded in the Taylor series as follows:

$$\psi_N = \psi_B + \frac{\partial \psi_B}{\partial y}\Delta y + \frac{1}{2}\frac{\partial^2 \psi}{\partial y^2}|_B \Delta y^2 + \frac{1}{6}\frac{\partial^3 \psi}{\partial y^3}|_B \Delta y^3 + O(\Delta^4). \quad (6.26)$$

If the wall is impervious to fluid, we see that $v = 0$ at B. It follows from (6.2) and (6.4) that

$$\omega_B = -\frac{\partial^2 \psi}{\partial y^2}|_B, \qquad \frac{\partial \omega_B}{\partial y} = -\frac{\partial^3 \psi}{\partial y^3}|_B.$$

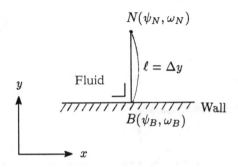

Figure 6.3: Boundary condition for vorticity.

Substituting these relations into (6.26) and neglecting the terms of higher order than third, we have

$$\psi_N = \psi_B + \frac{\partial \psi_B}{\partial y} \Delta y - \frac{\omega_B}{2} \Delta y^2 - \frac{1}{6} \frac{\partial \omega_B}{\partial y} \Delta y^3 \ .$$

The first-order derivative of the vorticity is approximated by the finite difference:

$$\frac{\partial \omega_B}{\partial y} \approx \frac{\omega_N - \omega_B}{\Delta y} \ .$$

In this case, we can see that the boundary value of ω is given by the formula:

$$\omega_{\text{wall}} = -\frac{3}{\ell^2} \left(\psi_N - \psi_B + \frac{\partial \psi_B}{\partial y} \ell \right) - \frac{\omega_N}{2} \ , \qquad (6.27)$$

where $\ell = \Delta y$, and the first-order derivative of the streamfunction implies that

$$\frac{\partial \psi_B}{\partial y} = \begin{cases} 0 & \text{on the non-slip boundary,} \\ V_s & \text{on the moving boundary,} \end{cases}$$

with some given tangential fluid velocity V_s.

6.4   Computational Scheme

The numerical solution is obtained step by step by using the discretised equations (6.22) and (6.23) in the following manner. With the initial condition $\omega^0 = 0$, which is assumed all over the domain for convenience' sake, values of the stream-function ψ^1 are calculated from (6.22). With the values ω^0 and ψ^1 thus obtained, new values of the vorticity ω^1 are calculated from (6.23). This procedure is continued for $n = 0, 1, 2, \ldots$ until the calculated flow fully develops. This iterative scheme is illustrated in Figure 6.4.

To examine the consistency and stability of the scheme, we consider the following convective diffusion equation in one dimension as a mathematical model

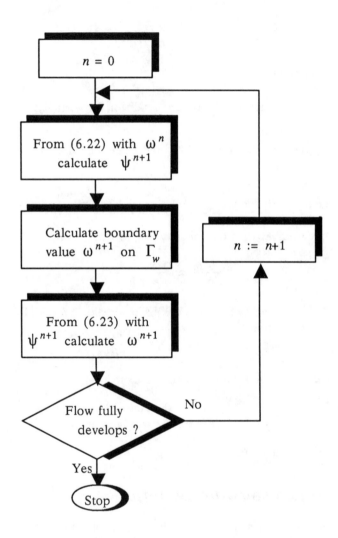

Figure 6.4: Computational scheme.

for the sake of simplicity.

$$\frac{\partial u}{\partial t} + v\frac{\partial u}{\partial x} = \lambda \frac{\partial^2 u}{\partial x^2}, \quad -\infty < x < +\infty, \quad t > 0 \tag{6.28}$$

with a given constant velocity v and a given positive constant λ.

Let x_j denote the nodal abscissae defined by $x_{j+1} = x_j + \Delta x$ $(j = 0, \pm1, \pm2, \ldots)$ with the spacing Δx. The finite element discretisation using the linear interpolation functions ϕ_α yields

$$\frac{1}{6}(\dot{u}_{j-1} + 4\dot{u}_j + \dot{u}_{j+1}) + v\frac{u_{j+1} - u_{j-1}}{2\Delta x}$$
$$= \lambda \frac{u_{j+1} - 2u_j + u_{j-1}}{\Delta x^2}, \tag{6.29}$$

where $u_j(t)$ denotes the approximate function to the exact $u(x_j, t)$ at the node x_j. Applying the Euler's first-order finite difference to the time derivatives involved in (6.29) for their approximations, we can obtain the following semi-implicit scheme.

$$\frac{1}{6}\left(\frac{u_{j-1}^{n+1} - u_{j-1}^n}{\Delta t} + 4\frac{u_j^{n+1} - u_j^n}{\Delta t} + \frac{u_{j+1}^{n+1} - u_{j+1}^n}{\Delta t}\right)$$
$$+ v\frac{u_{j+1}^n - u_{j-1}^n}{2\Delta x} = \lambda \frac{u_{j+1}^{n+1} - 2u_j^{n+1} + u_{j-1}^{n+1}}{\Delta x^2}, \tag{6.30}$$

where u_j^n denotes the approximate value to the exact $u(x_j, t_n)$.

We shall show that the scheme (6.30) is consistent with the differential equation (6.28). To this end, we assume the existence of a smooth function $U(x, t)$, which satisfies the difference equation:

$$\frac{1}{6}\left\{\frac{U(x_{j-1}, t_{n+1}) - U(x_{j-1}, t_n)}{\Delta t} + 4\frac{U(x_j, t_{n+1}) - U(x_j, t_n)}{\Delta t}\right.$$
$$\left. + \frac{U(x_{j+1}, t_{n+1}) - U(x_{j+1}, t_n)}{\Delta t}\right\} + v\frac{U(x_{j+1}, t_n) - U(x_{j-1}, t_n)}{2\Delta x}$$
$$= \lambda \frac{U(x_{j+1}, t_{n+1}) - 2U(x_j, t_{n+1}) + U(x_{j-1}, t_{n+1})}{\Delta x^2} \tag{6.31}$$

By expanding $U(x, t)$ into its Taylor series at (x_j, t_n), we can see that

$$\frac{U(x_{j+1}, t_n) - U(x_{j-1}, t_n)}{2\Delta x} = \frac{\partial U}{\partial x} + \frac{\Delta x^2}{6}\frac{\partial^3 U}{\partial x^3} + \frac{\Delta x^4}{120}\frac{\partial^5 U}{\partial x^5} + \cdots, \tag{6.32}$$

where the right-hand side has to be evaluated at (x_j, t_n). Substituting this expansion as well as (5.31) and (5.32) into (6.31), we obtain

$$\frac{\partial U}{\partial t} + v\frac{\partial U}{\partial x} = \lambda \frac{\partial^2 U}{\partial x^2} - \frac{\Delta t}{2}\frac{\partial^2 U}{\partial t^2} - v\frac{\Delta x^2}{6}\frac{\partial^3 U}{\partial x^3}$$
$$+ \left(\lambda\Delta t - \frac{\Delta x^2}{6}\right)\frac{\partial^3 U}{\partial x^2 \partial t} - \frac{\Delta t^2}{6}\frac{\partial^3 U}{\partial t^3} + \frac{\lambda}{12}\Delta x^2\frac{\partial^4 U}{\partial x^4}$$
$$+ \frac{\Delta t}{12}(6\lambda\Delta t - \Delta x^2)\frac{\partial^4 U}{\partial x^2 \partial t^2} - \frac{\Delta t^3}{24}\frac{\partial^4 U}{\partial t^4} + \cdots, \tag{6.33}$$

where the truncated terms contain derivatives with the factors of $\Delta x^\alpha \Delta t^\beta$ ($\alpha + \beta \geq 4$). From this result it can been seen that the scheme (6.30) is consistent.

We shall transform the time derivatives on the right-hand side of (6.33) to the corresponding spatial derivatives. For this purpose, we notice that

$$
\begin{aligned}
\frac{\partial}{\partial t} &= -v \frac{\partial}{\partial x} + \lambda \frac{\partial^2}{\partial x^2} - \frac{\Delta t}{2} \frac{\partial^2}{\partial t^2} + \cdots \\
&= -v \frac{\partial}{\partial x} + \lambda \frac{\partial^2}{\partial x^2} \\
&\quad - \frac{\Delta t}{2} \left(-v \frac{\partial}{\partial x} + \lambda \frac{\partial^2}{\partial x^2} - \frac{\Delta t}{2} \frac{\partial^2}{\partial t^2} + \cdots \right)^2 + \cdots \\
&= -v \frac{\partial}{\partial x} + \left(\lambda - \frac{\Delta t}{2} v^2 \right) \frac{\partial^2}{\partial x^2} + \cdots .
\end{aligned} \tag{6.34}
$$

Using this relation, we can rewrite (6.33) as follows.

$$
\begin{aligned}
\frac{\partial U}{\partial t} &+ v \frac{\partial U}{\partial x} = \lambda \frac{\partial^2 U}{\partial x^2} \\
&- \frac{\Delta t}{2} \left\{ -v \frac{\partial}{\partial x} + \left(\lambda - \frac{\Delta t}{2} v^2 \right) \frac{\partial^2}{\partial x^2} + \cdots \right\}^2 U - v \frac{\Delta x^2}{6} \frac{\partial^3 U}{\partial x^3} \\
&+ \left(\lambda \Delta t - \frac{\Delta x^2}{6} \right) \frac{\partial^2}{\partial x^2} \left\{ -v \frac{\partial}{\partial x} + \left(\lambda - \frac{\Delta t}{2} v^2 \right) \frac{\partial^2}{\partial x^2} + \cdots \right\} U \\
&- \frac{\Delta t^2}{6} \left\{ -v \frac{\partial}{\partial x} + \left(\lambda - \frac{\Delta t}{2} v^2 \right) \frac{\partial^2}{\partial x^2} + \cdots \right\}^3 U + \frac{\lambda}{12} \Delta x^2 \frac{\partial^4 U}{\partial x^4} \\
&+ \frac{\Delta t}{12} \left(6\lambda \Delta t - \Delta x^2 \right) \frac{\partial^2}{\partial x^2} \left\{ -v \frac{\partial}{\partial x} + \cdots \right\}^2 U \\
&\qquad - \frac{\Delta t^3}{24} \left\{ -v \frac{\partial}{\partial x} + \cdots \right\}^4 U + \cdots \\
&= \left(\lambda - \frac{\Delta t}{2} v^2 \right) \frac{\partial^2 U}{\partial x^2} - \frac{\Delta t^2}{3} v^3 \frac{\partial^3 U}{\partial x^3} \\
&\qquad + \frac{\Delta t}{2} \lambda^2 \left(1 - \frac{1}{6d} + \frac{c^4}{6d^2} \right) \frac{\partial^4 U}{\partial x^4} + \cdots , \tag{6.35}
\end{aligned}
$$

with two non-dimensional numbers called the *Courant number* $c = v\Delta t/\Delta x$ and the *diffusion number* $d = \lambda \Delta t/\Delta x^2$. By comparing (6.35) with (6.28), we can see that an extra factor of $-\Delta t v^2/2$ to the constant λ comes out in the coefficient of the second-order derivative. Since this term corresponds to the viscosity of the fluid, the factor is called an *artificial viscosity*. We notice that the scheme (6.30) gives rise to the negative artificial viscosity. When λ is replaced by $\lambda_a = \lambda + \Delta t v^2/2$, the artificial viscosity can be eliminated.

Moreover, the additional third- and fourth-order derivatives appear in (6.35), which are absent in the original equation (6.28). How do they act? To answer this question, we consider the equation

$$
\frac{\partial U}{\partial t} + v \frac{\partial U}{\partial x} = \lambda_a \frac{\partial^2 U}{\partial x^2} - \nu_2 \frac{\partial^3 U}{\partial x^3} - \mu_2 \frac{\partial^4 U}{\partial x^4} . \tag{6.36}
$$

Assume that the solution to this equation can be expressed by the Fourier transform:

$$U(x, t) = \int_{-\infty}^{+\infty} a(\sigma) e^{i(\omega t - \sigma x)} \, d\sigma .$$ (6.37)

From (6.36) it follows that

$$i\omega - i\sigma v = \lambda(-i\sigma)^2 - \nu_2(-i\sigma)^3 - \mu_2(-i\sigma)^4 .$$

Hence we have

$$i\omega = -(\lambda_a \sigma^2 + \mu_2 \sigma^4) + i(v\sigma - \nu_2 \sigma^3) ,$$

and the solution has the form

$$U(x, t) = \int_{-\infty}^{+\infty} e^{-(\lambda_a + \mu_2 \sigma^2)\sigma^2 t} a(\sigma) e^{i\{(v - \mu_2 \sigma^2)t - x\}\sigma} \, d\sigma .$$ (6.38)

From this expression, we know that a wave component with the wave number σ is propagated with the speed of $v - \mu_2 \sigma^2$ and that its amplitude is decreased at the rate of $Exp[-(\lambda_a + \mu_2 \sigma^2)\sigma^2 t]$ as long as $\lambda_a + \mu_2 \sigma^2 > 0$ $(-\infty < \sigma < \infty)$. The propagation speed is dependent on the frequency. On the other hand, any wave components travelling according to (6.28) propagate with the constant speed v. We can see that odd-order derivatives greater than or equal to the third order are *dispersive*. The wave components involved in (6.38) at the high frequencies are damped more strongly than those at the low frequencies. The damping factor corresponding to (6.28) is $Exp[-\lambda \sigma^2 t]$. We see that even-order derivatives are *dissipative*.

We shall show that the semi-implicit scheme (6.30) is conditionally stable. To this end, we assume that the solution of (6.31) can be expressed again in a form given by (6.37). From (6.31) we have

$$\frac{1}{6} \left\{ \frac{e^{i(\omega \Delta t + \sigma \Delta x)} - e^{i\sigma \Delta x}}{\Delta t} + 4 \frac{e^{i\omega \Delta t} - 1}{\Delta t} + \frac{e^{i(\omega \Delta t - \sigma \Delta x)} - e^{-i\sigma \Delta x}}{\Delta t} \right\}$$

$$+ v \frac{e^{-i\sigma \Delta x} - e^{i\sigma \Delta x}}{2\Delta x} = \lambda \frac{e^{i(\omega \Delta t - \sigma \Delta x)} - 2e^{i\omega \Delta t} + e^{i(\omega \Delta t + \sigma \Delta x)}}{\Delta x^2} .$$ (6.39)

The amplification factor is given by

$$e^{i\omega(\sigma)\Delta t} = \frac{2 + \cos(\sigma \Delta x) + 3ic \sin(\sigma \Delta x)}{2 + \cos(\sigma \Delta x) + 6d(1 - \cos(\sigma \Delta x))} .$$ (6.40)

The square of the absolute value of the complex number is given by

$$| e^{i\omega(\sigma)\Delta t} |^2 = \frac{\{2 + \cos(\sigma \Delta x)\}^2 + 9\{c \sin(\sigma \Delta x)\}^2}{\{2 + \cos(\sigma \Delta x) + 6d(1 - \cos(\sigma \Delta x))\}^2} .$$

Therefore, a sufficient condition for stability is

$$9\{c \sin(\sigma \Delta x)\}^2 \leq 12d\{2 + \cos(\sigma \Delta x)\}\{1 - \cos(\sigma \Delta x)\}$$
$$+ 36d^2 \{1 - \cos(\sigma \Delta x)\}^2 \quad \text{for all} \quad \sigma ,$$

which implies

$$3\,c^2\,\{1 + \cos(\sigma\Delta x)\} \leq 4\,d\,\{2 + \cos(\sigma\Delta x)\} + 12\,d^2\,\{1 - \cos(\sigma\Delta x)\}\,. \quad (6.41)$$

This inequality is fulfilled by the sufficiently small c. However, it seems difficult to derive a simple relation between c and d, satisfying the inequality for all σ. Here we impose the additional condition; $d \leq 1$ on (6.41). Then we can see that

$$3\,c^2\,\{1 + \cos(\sigma\Delta x)\} \leq 4\,d\,\{2 + \cos(\sigma\Delta x)\} + 12\,d\,\{1 - \cos(\sigma\Delta x)\}\,,$$

from which it follows that

$$\frac{3\,c^2}{4\,d} \leq \frac{7}{1 + \cos(\sigma\Delta x)} - 2\,.$$

The right-hand side is greater than or equal to 3/2 for $-1 \leq \cos(\sigma\Delta x) \leq 1$. This implies that

$$\frac{3\,c^2}{4\,d} \leq \frac{3}{2} \quad i.e., \quad \frac{c^2}{2d} \leq 1$$

with $d \leq 1$ is sufficient for the stability. In other words, we have

$$\Delta t \leq \min\{\frac{2\lambda}{v^2},\,\frac{\Delta x^2}{\lambda}\}\,. \quad (6.42)$$

Based on the above discussion, we use the orthotropic kinematic viscosities;

$$\nu_{ax} = \nu + \frac{1}{2}u^2\,\Delta t\,, \qquad \nu_{ay} = \nu + \frac{1}{2}v^2\,\Delta t\,, \quad (6.43)$$

in the x and y directions in the computation instead of the kinematic viscosity ν in order to reduce the artificial viscosity.

6.5   Numerical Examples

6.5.1   Flow facing a back step

We consider a two-dimensional transient flow of air over a back step according to Argyris et al. [1985]. Figure 6.5 shows the geometry and boundary conditions. The fluid is assumed to be initially at rest. A plugged flow at a constant velocity of $V = 10\ m/s$ in the x direction is prescribed along the inlet on the left-hand side. The ψ values on the inlet boundary are calculated as follows: From (6.4) we see

$$\psi(y) = \int_{0.02}^{y} u\,dy = \int_{0.02}^{y} V\,dy = 10\,(\,y - 0.02\,)\,,$$
$$\psi(0.056) = 0.36\,.$$

The material constants of air used in this example are; the density $\rho = 1.293\ kg/m^3$ and the viscosity $\mu = 1.73 \times 10^{-5}\ Pa \cdot s$. The corresponding

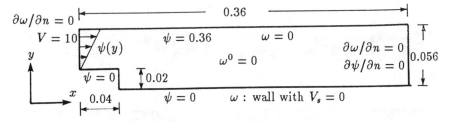

Figure 6.5: Geometry and boundary conditions.

kinematic viscosity is $\nu = 1.338 \times 10^{-5} \; m^2/s$. The Reynolds number, referenced to the step height as the characteristic length, is 14950. The outline of the geometry is defined by the combination of three blocks, as presented in Figure 6.6 (a). The mesh has 3000 triangular finite elements and 1601 nodes. Mesh grading in the relative ratio of 1:1:1:1:1:1:2:2:4:4 is used in the y direction in order to take the boundary layer effect into account. The interval $0 \le x \le 0.04$ is divided into 10 equal spacings and the interval $0.04 \le x \le 0.36$ is divided into 70 equal spacings. Time step is $\Delta t = 7.5 \times 10^{-5} \; s \; (= 75 \; \mu s)$. The evolution of the flow during 900 steps is calculated. This amounts merely to the final time of $0.0675 \; s \; (= 67.5 \; \mu s)$.

The calculated results are shown in Figure 6.7: The streamlines are presented at every 3.75 μs and every 7.5 μs.

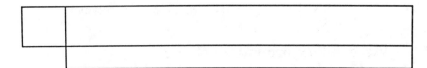

(a) Block mesh.

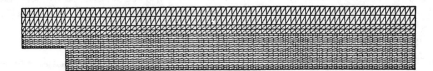

(b) Fine mesh.

Figure 6.6: Meshes.

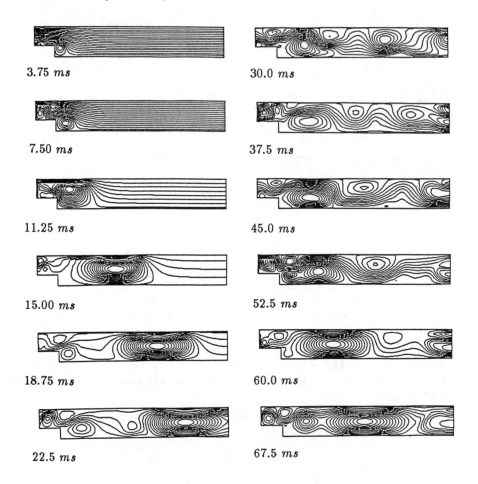

Figure 6.7: Calculated streamlines in a back step channel.

6.5.2   Viscous flow around a cylinder

We assume a uniform flow at a velocity of $u = 1$ m/s and $v = 0$ m/s, flowing from left to right toward a circular cylinder with radius $L = 1$ m, as illustrated in Figure 6.8 (a). The kinematic viscosity is assumed to be $\nu = 0.01$ m^2/s. Using the characteristic velocity $V = 1.0$ m/s and the length L, we see that $Re = 100$. The fluid is assumed to be initially motionless. The finite element mesh is presented in Figure 6.8 (b). The time increment, $\Delta t = 0.1$ s is used.

The calculated results are shown in Figure 6.9. The cylinder sheds the Kármán vorteces periodically. Let f denote the frequency of the Kármán vortex shedding $(1/s)$. The period for a circular cylinder with the diameter d is given theoretically by $1/f = d/(0.2V)$. The period obtained from this numerical example is 10 s, which agrees with the theoretical value. The Strouhal number in this case is $St = fd/V = 0.2$. The movement of particles can be traced along with the fluid motion using the Euler's method. The calculated distribution of

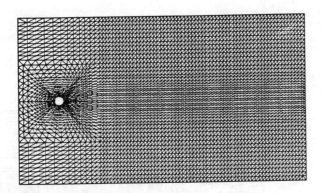

$\psi_B = 20$, $\omega_B = 0$

$1\,m/s$

$1\,m$

$\psi_B = y$ $\psi_B = 0$

$\omega_B = 0$ $\omega = \omega_{\text{wall}}$

$\dfrac{\partial \psi}{\partial n}\Big|_B = 0$

$\dfrac{\partial \omega}{\partial n}\Big|_B = 0$

$40\,m$

y $\psi_B = -20$, $\omega_B = 0$

$\vdash$ 10 $\dashv\vdash$ 60

x

(a) Geometry and boundary conditions.

(b) Finite element mesh.

Figure 6.8: Viscous flow around a cylinder.

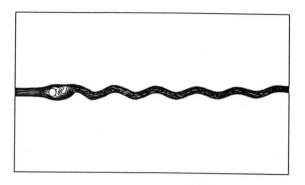

(a) Streamlines ($-1 \leq \psi \leq 1$).

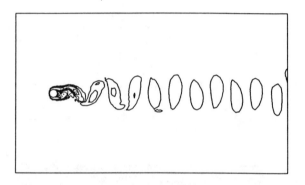

(b) Vorticity contours ($-1.64 \leq \omega \leq 1.64$).

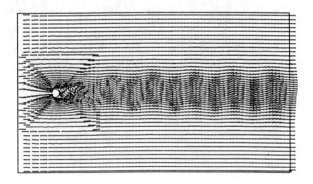

(c) Flow vectors (maximum velocity $1.35\,m/s$).

Figure 6.9: Calculated viscous flow behind a circular cylinder; $Re = 100, t = 120s$.

the particles at $t = 170$ s is presented in the picture on the front cover of this book.

6.5.3   Cavity flow in a rectangle

We consider a wind-driven cavity flow in two dimensions at Reynolds numbers $Re = 1000$ and 10000, as shown in Figure 6.10(a). The fluid is assumed to be initially motionless. The top surface is forced to move horizontally from the left to right at a velocity of $U = 1$ m/s. We used $L = 1$ m, $\nu = 0.001$ and 0.0001 m^2/s respectively, and $\Delta t = 0.1$ s.

The calculated results, when the flow fully develops, are presented in Figure 6.11. According to Ghia et $al.$ [1982], the center of the main vortex at $Re = 1000$ has the coordinates (0.5313, 0.5625). The secondary vortices are observed near the two bottom corners. According to Ghia et $al.$, the center of the right secondary vortex has the coordinates (0.8906, 0.1250). Velocity profiles of u along the vertical center line are plotted in Figure 6.12.

6.5.4   Rates of Convergence

We shall briefly examine the rates of convergence in the finite element approximation of the streamfunction and vorticity. To make the problem mathematically tractable, we confine our discussion to the following problem:

$$-\nabla\psi = \omega \quad \text{in} \quad \Omega\,, \tag{6.44}$$

$$\frac{\partial\omega}{\partial t} + J(\omega, \psi) = \nu\nabla\omega + f \quad \text{in} \quad \Omega\,, \tag{6.45}$$

subject to the homogeneous initial and boundary conditions:

$$\psi = 0 \quad \text{in} \quad \Omega\,, \quad \text{at} \quad t = 0\,,$$

$$\psi = \frac{\partial\psi}{\partial n} = 0 \quad \text{on} \quad \Gamma\,. \tag{6.46}$$

Here $J(\omega, \psi)$ is the Jacobian which represents the convective terms, and f denotes a body force which has been incorporated from any inhomogeneous initial boundary conditions. We assume that Ω is enclosed by a smooth Jordan curve Γ in the plane. In this section, we shall briefly review the results due to G. Fix [1988], without the proofs.

The problem in the weak form corresponding to the problems (6.44)-(6.46) consists of finding the unknown ψ in $H_0^1(\Omega)$ and the unknown ω in $H^1(\Omega)$, satisfying

$$(\nabla\psi, \nabla\chi) = (\omega, \chi) \quad \text{for} \quad \forall\chi \in H^1(\Omega)\,, \tag{6.47}$$

$$(\frac{\partial\omega}{\partial t}, \zeta) + (J(\omega, \psi), \zeta) + \nu(\nabla\omega, \nabla\zeta) = (f, \zeta) \tag{6.48}$$

$$\text{for} \quad \forall\zeta \in H_0^1(\Omega)\,,$$

where $(\cdot, \cdot)$ denotes the $L^2(\Omega)$-scalar product, $H^1(\Omega)$ denotes the conventional Sobolev space, and $H_0^1(\Omega)$ is the subspace of $H^1(\Omega)$ with zero boundary values.

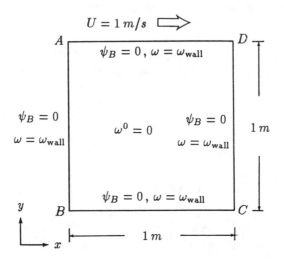

(a) Boundary conditions.

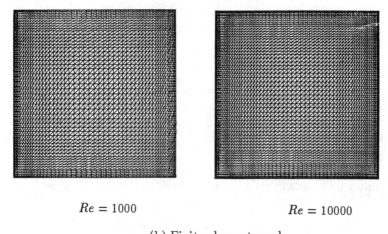

$Re = 1000$ $Re = 10000$

(b) Finite element mesh.

Figure 6.10: Wind-driven cavity flow.

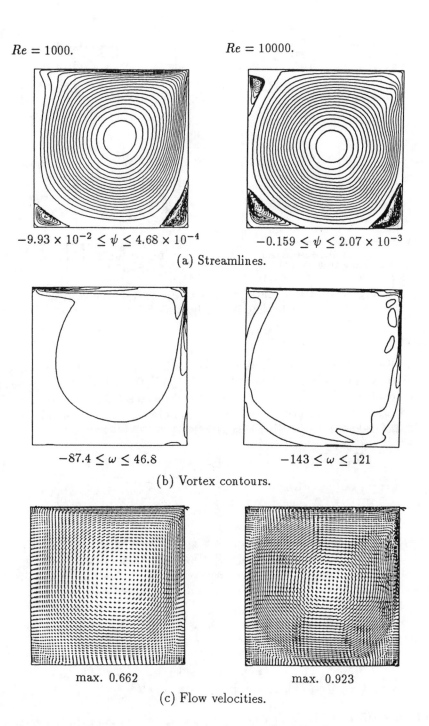

$Re = 1000.$

$Re = 10000.$

$-9.93 \times 10^{-2} \leq \psi \leq 4.68 \times 10^{-4}$ $-0.159 \leq \psi \leq 2.07 \times 10^{-3}$

(a) Streamlines.

$-87.4 \leq \omega \leq 46.8$ $-143 \leq \omega \leq 121$

(b) Vortex contours.

max. 0.662 max. 0.923

(c) Flow velocities.

Figure 6.11: Calculated wind-driven cavity flows.

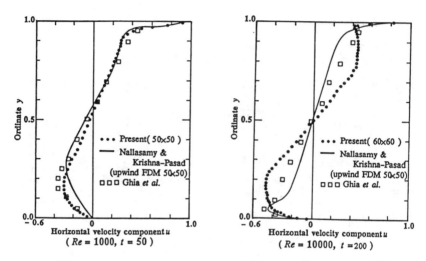

Figure 6.12: Horizontal velocity component on the vertical
center line in steady state.

As discussed in Section 5.5, we consider again the regular triangulation of Ω
and linear triangular finite elements. Let $\mathcal{S}_h \subset H^1(\Omega)$ and $\mathcal{S}_0^h \subset H_0^1(\Omega)$ be the
corresponding finite element subspaces. When we recast the weak forms (6.47)
and (6.48) in these subspaces, we obtain a semi-discrete problem of unknown ψ_h
in $\mathcal{S}_0^h$ and unknown ω_h in $\mathcal{S}^h$, which satisfy

$$(\nabla\psi_h, \nabla\chi_h) = (\omega_h, \chi_h) \quad \text{for} \quad \forall\chi_h \in \mathcal{S}^h, \tag{6.49}$$

$$\left(\frac{\partial\omega_h}{\partial t}, \zeta_h\right) + (J(\omega_h, \psi_h), \zeta_h) + \nu(\nabla\omega_h, \nabla\zeta_h) = (f, \zeta_h) \tag{6.50}$$

$$\text{for} \quad \forall\zeta_h \in \mathcal{S}_0^h.$$

We see that our approximate method is a *mixed formulation*.

Suppose that the problem is a steady state one. Then the term, in which the
time derivative is involved, vanishes, and we have the following error estimates
in the L^2-norm.

Theorem 6.1 *The errors in $\nabla\psi_h$ and ω_h using the linear triangular finite elements
in the steady-state flow problems are bounded by*

$$\|\nabla\psi_h - \nabla\psi\| \leq C_1(\psi, \omega)\, h, \tag{6.51}$$

$$\|\omega_h - \omega\| \leq C_2(\psi, \omega)\, h^{1/2}, \tag{6.52}$$

with the positive constants C_1 and C_2, depending only on the variables specified.

From (6.51) we can see that the rate of convergence in the velocity vectors is optimal. However, from (6.52) we see that the rate of convergence in the vorticity field is sub-optimal.

Exercises

6.1 Consider a two-dimensional viscous fluid flow with the parabolic velocity
profile in a channel having a cylindrical obstacle on its bed, as shown in
Figure 6.13. The inlet flow profile is given as follows.

$$u = -0.048\,y^2 + 0.48\,y\,, \quad v = 0\,,$$

$$\psi = \int_0^y u(y)\,dy = -0.016\,y^3 + 0.24\,y^2\,,$$

$$\omega = -\frac{du}{dy} = 0.096\,y - 0.48\,.$$

Take $\nu = 0.05\ m^2/s$ and $\Delta t = 0.2\ s$. Along the side AB in Figure 6.13(a),

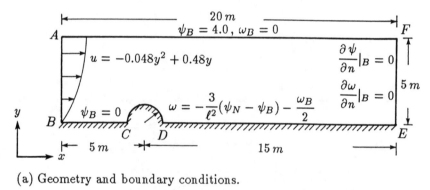

(a) Geometry and boundary conditions.

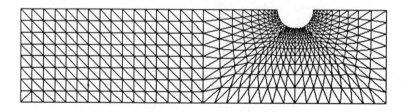

(b) Finite element mesh.

Figure 6.13: Parabolic flow over a half cylindrical obstacle.

the boundary values at the corresponding nodal points in Figure 6.13(b)
are summarized in Table 6.1. Assume that the fluid is initially motionless.
Calculate the flow until it fully develops. Compare your results with the
calculated results presented in Figure 6.14.

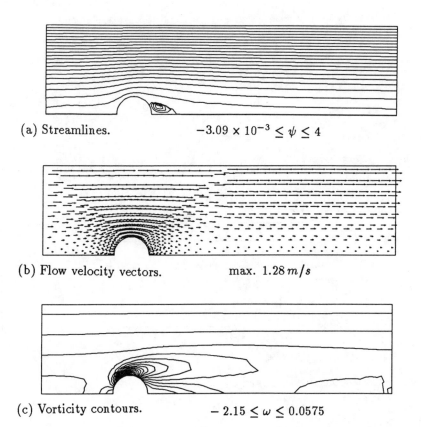

(a) Streamlines. $-3.09 \times 10^{-3} \leq \psi \leq 4$

(b) Flow velocity vectors.   max. $1.28\,m/s$

(c) Vorticity contours. $-2.15 \leq \omega \leq 0.0575$

Figure 6.14: Calculated flow over an obstacle.

Table 6.1: Boundary values on the inlet.

$y\,(m)$	$u\,(m/s)$	$\psi\,(m^2/s)$	$\omega\,(1/s)$
0.00	0.00	0.00	-0.48
0.625	0.28125	0.0898	-0.42
1.25	0.525	0.3438	-0.36
1.875	0.73125	0.7383	-0.30
2.50	0.9	1.25	-0.24
3.125	1.03125	1.8555	-0.18
3.75	1.125	2.5313	-0.12
4.375	1.18125	3.253	-0.06
5.00	1.20	4.00	0.00

Chapter 7

THERMAL FLUID FLOW

Chapter 5 dealt with the finite element analysis of the heat conduction in a solid. Chapter 6 dealt with the finite element flow analysis of a viscous fluid. In this chapter, we shall consider an extension of those analyses to the heat transfer in the viscous fluid. A common fluid is a heat-conductive medium. However, when the temperature difference becomes large enough, thermal energy is transported simultaneously by conduction and convection processes. In the problems of thermal fluid flow, the equation of thermal energy is coupled with the equations of viscous fluid flow. As numerical examples, we shall present the natural convection in a closed compartment with a heated side wall, and in an open vessel heated from below as well as the forced thermal convection.

7.1 Governing Equations

As we have seen in the preceding chapters, the basic equations involve various physical constants. For example, in (5.3) and (6.1) included are the density, viscosity, heat conduction coefficient, *etc.* The values of these constants generally depend on the temperature. Here we propose an assumption, called the *Boussinesq approximation*, this is a physical assumption, saying that all the coefficients are independent of the temperature except a term which describes the buoyancy F_y per unit mass in the equations of motion:

$$\frac{\partial u}{\partial t} + u\frac{\partial u}{\partial x} + v\frac{\partial u}{\partial y} = -\frac{1}{\rho}\frac{\partial p}{\partial x} + \nu\nabla^2 u , \qquad (7.1)$$

$$\frac{\partial v}{\partial t} + u\frac{\partial v}{\partial x} + v\frac{\partial v}{\partial y} = -\frac{1}{\rho}\frac{\partial p}{\partial y} + \nu\nabla^2 v + F_y(T) . \qquad (7.2)$$

We shall derive the expression of F_y, depending on the temperature T. To this end, we consider a fluid at temperature T_0 and density ρ_0 in a container, as shown in Figure 7.1. The ordinate y is taken in the opposite direction to the gravitational acceleration g. We assume that a lump of the fluid in the container is heated or cooled to the temperature T with its density ρ. Then, the *buoyancy* of magnitude $(\rho_0 - \rho)g \ (N/m^3)$ is exerted on the lump of the fluid. If the density difference is small enough, the equation of state of the fluid can be expressed as follows.

$$\frac{\rho}{\rho_0} = 1 - \beta(T - T_0),\tag{7.3}$$

where $\beta \ (1/K)$ is the *volumetric thermal expansion coefficient*. The buoyancy per a unit mass is therefore given by

$$F_y = (\rho_0 - \rho)g/\rho_0 = \beta g(T - T_0).\tag{7.4}$$

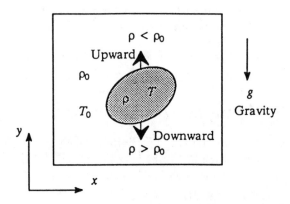

Figure 7.1: Buoyancy induced by non-uniform temperature.

The continuity equation remains unchanged as was given by (6.3) under the Boussinesq approximation:

$$\frac{\partial u}{\partial x} + \frac{\partial v}{\partial y} = 0.\tag{7.5}$$

With the buoyancy given by (7.4), the basic equations expressing the fluid flow are written in terms of the streamfunction and vorticity as follows.

$$\nabla^2 \psi = -\omega,\tag{7.6}$$

$$\frac{\partial \omega}{\partial t} + \frac{\partial \psi}{\partial y}\frac{\partial \omega}{\partial x} - \frac{\partial \psi}{\partial x}\frac{\partial \omega}{\partial y} = \nu \nabla^2 \omega + g\beta\frac{\partial T}{\partial x}.\tag{7.7}$$

Boundary conditions concerning ψ and ω are given as in Figure 6.1 and as discussed in Section 6.3.

The thermal energy is transported simultaneously by conduction and convection. When the dissipation of kinematic energy to heat due to the viscosity is neglected, the components of the heat flux are given by

$$q_x = -\kappa \frac{\partial T}{\partial x} + u\rho cT, \tag{7.8}$$

$$q_y = -\kappa \frac{\partial T}{\partial y} + v\rho cT. \tag{7.9}$$

From the law of conservation of thermal energy described as

$$\rho c \frac{\partial T}{\partial t} + \frac{\partial q_x}{\partial x} + \frac{\partial q_y}{\partial y} = 0, \tag{7.10}$$

we can obtain the following convective heat conduction equation.

$$\frac{\partial T}{\partial t} + \frac{\partial \psi}{\partial y}\frac{\partial T}{\partial x} - \frac{\partial \psi}{\partial x}\frac{\partial T}{\partial y} = \lambda \nabla^2 T \tag{7.11}$$

with the thermal conductivity $\lambda = \kappa/(\rho c)$. Boundary conditions concerning T are given by (5.5)-(5.7).

7.2   Finite Element Discretisation

We shall apply the conventional Galerkin finite element method to discretising of the streamfunction equation (7.6), the vorticity transport equation (7.7), and the convective heat conduction equation (7.11). To this end, let $\delta\psi$, $\delta\omega$ and δT be weighting functions, being arbitrary but $\delta\psi = 0$ on Γ_ψ, $\delta\omega = 0$ on $\Gamma_\omega \cup \Gamma_w$, and $\delta T = 0$ on Γ_T. We start the finite element formulation with the weighted residual forms of those equations:

$$\int_\Omega \delta\psi \, \nabla^2\psi \, d\omega + \int_\Omega \delta\psi \, \omega \, d\Omega = 0, \tag{7.12}$$

$$\int_\Omega \delta\omega \frac{\partial \omega}{\partial t} \, d\Omega + \int_\Omega \delta\omega \left(\frac{\partial \psi}{\partial y}\frac{\partial \omega}{\partial x} - \frac{\partial \psi}{\partial x}\frac{\partial \omega}{\partial y} \right) d\Omega$$
$$- \int_\Omega \delta\omega \, g\beta \frac{\partial T}{\partial x} \, d\Omega - \int_\Omega \delta\omega \, \nu \nabla^2\omega \, d\Omega = 0, \tag{7.13}$$

$$\int_\Omega \delta T \frac{\partial T}{\partial t} \, d\Omega + \int_\Omega \delta T \left(\frac{\partial \psi}{\partial y}\frac{\partial T}{\partial x} - \frac{\partial \psi}{\partial x}\frac{\partial T}{\partial y} \right) d\Omega$$
$$- \int_\Omega \delta T \, \lambda \nabla^2 T \, d\Omega = 0. \tag{7.14}$$

Integration by parts of the terms involving the Laplacian yields

$$\int_\Omega \nabla\delta\psi \cdot \nabla\psi \, d\Omega - \int_\Omega \delta\psi \, \omega \, d\Omega - \int_{\Gamma_\bullet} \delta\psi \frac{\partial \psi}{\partial n} \, d\Gamma = 0. \tag{7.15}$$

$$\int_\Omega \delta\omega \frac{\partial \omega}{\partial t} \, d\Omega + \int_\Omega \delta\omega \left(\frac{\partial \psi}{\partial y}\frac{\partial \omega}{\partial x} - \frac{\partial \psi}{\partial x}\frac{\partial \omega}{\partial y} \right) d\Omega + \int_\Omega \nu \nabla\delta\omega \cdot \nabla\omega \, d\Omega$$

$$- \int_\Omega \delta\omega \, g \, \beta \frac{\partial T}{\partial x} \, d\Omega - \int_{\Gamma_x} \nu \, \delta\omega \frac{\partial \omega}{\partial n} \, d\Gamma = 0 , \qquad (7.16)$$

$$\int_\Omega \delta T \frac{\partial T}{\partial t} \, d\Omega + \int_\Omega \delta T \left(\frac{\partial \psi}{\partial y} \frac{\partial T}{\partial x} - \frac{\partial \psi}{\partial x} \frac{\partial T}{\partial y} \right) d\Omega + \int_\Omega \lambda \, \nabla \delta T \cdot \nabla T \, d\Omega$$

$$- \int_{\Gamma_q} Q_B \, \delta T \, d\Gamma - \int_{\Gamma_h} Q_h(T) \, \delta T \, d\Gamma = 0 , \qquad (7.17)$$

where $Q_B = q_B/(\rho c)$ and $Q_h = h(T - T_a)/(\rho c)$.

The domain Ω is divided into triangular finite elements. Inside each triangle e having its three vertices as the element nodes with their local node numbers 1, 2, 3; the unknown streamfunction, vorticity and temperature are linearly interpolated as follows.

$$\psi = \sum_\alpha \phi_\alpha \psi_\alpha , \qquad \delta\psi = \sum_\alpha \phi_\alpha \delta\psi_\alpha ,$$

$$\omega = \sum_\alpha \phi_\alpha \omega_\alpha , \qquad \delta\omega = \sum_\alpha \phi_\alpha \delta\omega_\alpha , \qquad (7.18)$$

$$T = \sum_\alpha \phi_\alpha T_\alpha , \qquad \delta T = \sum_\alpha \phi_\alpha \delta T_\alpha ,$$

where the notation $\sum_\alpha$ indicates taking sums for all α ($\alpha = 1, 2, 3$), and ϕ_α are linear interpolation functions given by

$$\phi_\alpha = \frac{1}{2 \, \Delta^e} (a_\alpha + b_\alpha x + c_\alpha y) \qquad (7.19)$$

with the area of the triangle Δ^e; $\psi_\alpha, \omega_\alpha, T_\alpha$ are nodal values of the corresponding unknowns; and $\delta\psi_\alpha, \delta\omega_\alpha, \delta T_\alpha$ are their arbitrary nodal variations.

The interpolations (7.18) are substituted into (7.15)-(7.17). From the continuity in the interpolations of ψ, ω, T and the arbitrariness of $\delta\psi_\alpha, \delta\omega_\alpha, \delta T_\alpha$, the next element equations follow.

$$\sum_{\beta=1}^{3} D_{\alpha\beta}^e \psi_\beta - \sum_{\beta=1}^{3} M_{\alpha\beta}^e \omega_\beta - \Gamma_{s\alpha}^e = 0 , \qquad (7.20)$$

$$\sum_{\beta=1}^{3} M_{\alpha\beta}^e \dot\omega_\beta + \sum_{\beta=1}^{3} A_{\alpha\beta}^e \omega_\beta + \nu \sum_{\beta=1}^{3} D_{\alpha\beta}^e \omega_\beta$$

$$- F_\alpha^e(T) - \Gamma_{\chi\alpha}^e = 0 , \qquad (7.21)$$

$$\sum_{\beta=1}^{3} M_{\alpha\beta}^e \dot T_\beta + \sum_{\beta=1}^{3} A_{\alpha\beta}^e T_\beta + \lambda \sum_{\beta=1}^{3} D_{\alpha\beta}^e T_\beta$$

$$+ \Gamma_{q\alpha}^e + \Gamma_{h\alpha}^e(T) = 0 , \qquad (7.22)$$

where the coefficients are given by

$$M_{\alpha\beta}^e = \int_e \phi_\alpha \phi_\beta \, d\Omega = \frac{\Delta^e}{12} (1 + \delta_{\alpha\beta}) , \qquad (7.23)$$

$$D_{\alpha\beta}^e = \int_e \left(\frac{\partial \phi_\alpha}{\partial x} \frac{\partial \phi_\beta}{\partial x} + \frac{\partial \phi_\alpha}{\partial y} \frac{\partial \phi_\beta}{\partial y} \right) \delta\Omega$$

$$= \frac{1}{4\Delta^e} (b_\alpha b_\beta + c_\alpha c_\beta) , \tag{7.24}$$

$$A_{\alpha\beta}^e = \int_e \phi_\alpha \left(\sum_{\gamma=1}^{3} \frac{\partial \phi_\gamma}{\partial y} \psi_\gamma \frac{\partial \phi_\beta}{\partial x} - \sum_{\gamma=1}^{3} \frac{\partial \phi_\gamma}{\partial x} \psi_\gamma \frac{\partial \phi_\beta}{\partial y} \right) d\Omega$$

$$= \frac{1}{12\Delta^e} \sum_{\gamma=1}^{3} (c_\gamma b_\beta - b_\gamma c_\beta) \psi_\gamma , \tag{7.25}$$

$$F_\alpha^e = g\beta \int_e \phi_\alpha \frac{\partial}{\partial x} \left(\sum_{\gamma=1}^{3} \phi_\gamma T_\gamma \right) d\Omega = \frac{g\beta}{6} \sum_{\gamma=1}^{3} b_\gamma T_\gamma , \tag{7.26}$$

$$\Gamma_{s\alpha}^e = \int_{\Gamma_s^e} \phi_\alpha \frac{\partial \psi}{\partial n} d\Gamma = -\frac{V_s}{2} \mid \Gamma_s^e \mid \begin{bmatrix} 0 \\ 1 \\ 1 \end{bmatrix} , \tag{7.27}$$

$$\Gamma_{\chi\alpha}^e = \int_{\Gamma_\chi^e} \nu \phi_\alpha \frac{\partial \omega}{\partial n} d\Gamma = \nu \frac{\chi_n}{2} \mid \Gamma_\chi^e \mid \begin{bmatrix} 0 \\ 1 \\ 1 \end{bmatrix} , \tag{7.28}$$

$$\Gamma_{q\alpha}^e = \int_{\Gamma_q^e} \phi_\alpha Q_B \, d\Gamma = \frac{Q_B}{2} \mid \Gamma_q^e \mid \begin{bmatrix} 0 \\ 1 \\ 1 \end{bmatrix} , \tag{7.29}$$

$$\Gamma_{h\alpha}^e = \int_{\Gamma_h^e} \phi_\alpha Q_h \, d\Gamma = \frac{Q_h}{2} \mid \Gamma_h^e \mid \begin{bmatrix} 0 \\ 1 \\ 1 \end{bmatrix} , \tag{7.30}$$

with $\Gamma_s^e = \Gamma_s \cap \partial e, \Gamma_\chi^e = \Gamma_\chi \cap \partial e, \Gamma_q^e = \Gamma_q \cap \partial e$, and $\Gamma_h^e = \Gamma_h \cap \partial e$ under a similar situation to the one in Figure 3.8.

After assembling all the element equations over the whole domain, we will obtain the total equations in the form:

$$[D]\{\psi\} - [M]\{\omega\} - \{\Gamma_s\} = \{0\} , \tag{7.31}$$

$$[M]\{\dot{\omega}\} + [A(\psi)]\{\omega\} + \nu[D]\{\omega\}$$
$$- \{F(T)\} - \{\Gamma_\chi\} = \{0\} , \tag{7.32}$$

$$[M]\{\dot{T}\} + [A(\psi)]\{T\} + \lambda[D]\{T\}$$
$$+ \{\Gamma_q\} + \{\Gamma_h(T)\} = \{0\} , \tag{7.33}$$

where $[D]$, $[M]$, $[A]$ denote the total matrices, $\{F\}$ corresponds the buoyancy term; $\{\psi\}, \{\omega\}, \{T\}$ are the nodal unknown column vectors. Other column vectors are due to the respective boundary conditions. One can read (7.31)-(7.33) as a nonlinear system of first-order ordinary differential equations for unknown $\{\psi\}, \{\omega\}$ and $\{T\}$.

7.3   Computational Scheme

We shall consider the discretisation of the total equations with respect to time. To this end, we apply the semi-implicit scheme to the initial value problem. The time derivatives of the nodal vorticity and temperature are approximated by the following finite differences,

$$\dot{\omega}_\beta = \frac{d\omega_\beta}{dt} \approx \frac{\omega_\beta^{n+1} - \omega_\beta^n}{\Delta t}, \tag{7.34}$$

$$\dot{T}_\beta = \frac{dT_\beta}{dt} \approx \frac{T_\beta^{n+1} - T_\beta^n}{\Delta t}, \tag{7.35}$$

where ω_β^n and T_β^n denote the nodal vorticity and temperature, respectively, at the time level t_n, defined by $t_{n+1} = t_n + \Delta t$ $(n = 0, 1, 2, ...)$ with the time increment Δt.

We consider the approximation of (7.31)-(7.33) at the time level t_{n+1} by replacing the time derivatives with (7.34) and (7.35). This results in

$$[D]\{\psi^{n+1}\} = [M]\{\omega_n\} + \{\Gamma_*^n\}, \tag{7.36}$$

$$\frac{1}{\Delta t}[M]\{\omega^{n+1}\} + \nu[D]\{\omega^{n+1}\}$$
$$= \frac{1}{\Delta t}[M]\{\omega^n\} - [A(\psi^{n+1})]\{\omega^n\}$$
$$+ \{F(T^n)\} + \{\Gamma_x^{n+1}\}, \tag{7.37}$$

$$\frac{1}{\Delta t}[M]\{T^{n+1}\} + \lambda[D]\{T^{n+1}\}$$
$$= \frac{1}{\Delta t}[M]\{T^n\} - [A(\psi^{n+1})]\{T^n\}$$
$$- \{\Gamma_q^{n+1}\} - \{\Gamma_h(T^n)\}. \tag{7.38}$$

Initially zero vorticity is assumed for most of the cases. Then, the computed $\{\psi_1\}$ using (7.36) corresponds the potential flow of the problem. In equation (7.38), the radiation term on the boundary Γ_h is evaluated at time t_n so that the following nonlinear radiation of the Stefan-Boltzmann type may also be considered.

$$- \kappa \frac{\partial T}{\partial n} = \sigma E(T^4 - T_r^4) \quad \text{on} \quad \Gamma_\sigma, \tag{7.39}$$

where σ $(W/m^2 \cdot K^4)$ is the Stefan-Boltzmann constant of the boundary surface Γ_σ, and E $(-)$ is given by the expression

$$E = V / (\frac{1}{\varepsilon} + \frac{1}{\varepsilon_r} - 1)$$

with the radiation view factor V, the emissivities ε and ε_r of the surface at the temperature T (K) and of the external radiating source at the temperature T_r (K), respectively.

The artificial kinematic viscosity as well as the artificial thermal conductivity are introduced in the computation. Instead of ν and λ, we use

$$\nu_{ax} = \nu + \frac{1}{2}u^2\,\Delta t \quad, \quad \nu_{ay} = \nu + \frac{1}{2}v^2\,\Delta t , \tag{7.40}$$

$$\lambda_{ax} = \lambda + \frac{1}{2}u^2\,\Delta t \quad, \quad \lambda_{ay} = \lambda + \frac{1}{2}v^2\,\Delta t . \tag{7.41}$$

In terms of the Algol-like statements, the computation proceeds as follows.

Read *topological data.*

Read *material constants and boundary conditions.*

Set initial values.

For $n = 0, 1, 2, \ldots$, **until** *satisfied*, **do:**

    Calculate $\{\psi^{n+1}\}$.

    Insert the boundary values ω_{wall} *on* Γ_w .

    Calculate $\{\omega^{n+1}\}$.

    Calculate $\{T^{n+1}\}$.

The algorithm is not complete until some stopping criteria are specified for the iteration counter n. When flow duration is given, n runs up to the integer $N = (t_f - t_0)/\Delta t$ with the final time t_f. Otherwise, the iteration may continue until the calculated thermal fluid motion fully develops.

We shall consider non-dimensional forms of the governing equations. For the *natural* or *free convection*, we shall introduce the following non-dimensional variables:

$$x^* = \frac{x}{L} , \qquad y^* = \frac{y}{L} , \qquad t^* = \frac{t}{L^2/\nu} ,$$

$$\tag{7.42}$$

$$u^* = \frac{u}{\nu/L} , \qquad v^* = \frac{v}{\nu/L} , \qquad T^* = \frac{T - T^0}{\Delta T}$$

with the characteristic length L and the characteristic temperature difference ΔT. Here, the quotient ν/L is chosen to represent a reference velocity. Corresponding to these variable transformations, non-dimensional streamfunction and vorticity are expressed in the form:

$$\psi^* = \frac{\psi}{\nu} , \qquad \omega^* = \frac{\omega}{\nu/L^2} . \tag{7.43}$$

From (7.6), (7.7) and (7.11), we can obtain the non-dimensional equations:

$$\nabla^{*2}\psi^* = -\omega^* , \tag{7.44}$$

$$\frac{\partial \omega^*}{\partial t^*} + \frac{\partial \psi^*}{\partial y^*}\frac{\partial \omega^*}{\partial x^*} - \frac{\partial \psi^*}{\partial x^*}\frac{\partial \omega^*}{\partial y^*} = \nabla^{*2}\omega^* + Gr\frac{\partial T^*}{\partial x^*} , \tag{7.45}$$

$$\frac{\partial T^*}{\partial t^*} + \frac{\partial \psi^*}{\partial y^*}\frac{\partial T^*}{\partial x^*} - \frac{\partial \psi^*}{\partial x^*}\frac{\partial T^*}{\partial y^*} = \frac{1}{Pr}\nabla^{*2}T^* , \tag{7.46}$$

with the Grashof number Gr given by (1.10) and the Prandtl number Pr given by (1.11). Only these two dimensionless numbers concern the similarity in the free convection.

For the *forced convection*, we shall introduce the following non-dimensional variables:

$$x^* = \frac{x}{L}, \qquad y^* = \frac{y}{L}, \qquad t^* = \frac{t}{L/U},$$

$$u^* = \frac{u}{U}, \qquad v^* = \frac{v}{U}, \qquad T^* = \frac{T - T^0}{\Delta T} \qquad (7.47)$$

with the characteristic velocity U. Corresponding to these transformations, non-dimensional streamfunction and vorticity are expressed in the form:

$$\psi^* = \frac{\psi}{LU}, \qquad \omega^* = \frac{\omega}{U/L}. \qquad (7.48)$$

From (7.6), (7.7) and (7.11), we can obtain the non-dimensional equations:

$$\nabla^{*2}\psi^* = -\omega^*, \qquad (7.49)$$

$$\frac{\partial \omega^*}{\partial t^*} + \frac{\partial \psi^*}{\partial y^*}\frac{\partial \omega^*}{\partial x^*} - \frac{\partial \psi^*}{\partial x^*}\frac{\partial \omega^*}{\partial y^*} = \frac{1}{Re}\nabla^{*2}\omega^* + \frac{1}{Fr}\frac{\partial T^*}{\partial x^*}, \qquad (7.50)$$

$$\frac{\partial T^*}{\partial t^*} + \frac{\partial \psi^*}{\partial y^*}\frac{\partial T^*}{\partial x^*} - \frac{\partial \psi^*}{\partial x^*}\frac{\partial T^*}{\partial y^*} = \frac{1}{Re\,Pr}\nabla^{*2}T^*, \qquad (7.51)$$

with the *Froude number* Fr defined by

$$\frac{1}{Fr} = \frac{g\,\beta\,\Delta T\,L}{U^2} = \frac{Gr}{Re^2}. \qquad (7.52)$$

We see that three dimensionless numbers Re, Fr, Pr concern the similarity in the forced thermal convection.

7.4   Numerical Examples

7.4.1   Natural convection in a closed compartment

We consider a compartment with rectangular cross section, as shown in Figure 7.2. The water in the compartment is assumed to be initially motionless and at a temperature of 21.5 °C. The left side wall is cooled at a temperature of 20.0 °C and the right side wall is heated at 23 °C, so that the compartment keeps the temperature difference of $\Delta T = 3.0$ °C. The top and bottom are fixed in their positions and are thermally insulated. The boundary conditions are indicated in Figure 7.2.

Physical constants of water used in this example are; $\rho = 999.8 \; kg/m^3$, $\mu = 1.793 \times 10^{-3} \; Pas$, $c = 4217 \; J/kg \cdot K$, $\kappa = 0.5683 \; W/m \cdot K$, $\beta = 4.6 \times 10^{-4} \; 1/K$, $g = 9.81 \; m/s^2$. We used the time increment $\Delta t = 0.1 \; s$. With the characteristic length of the compartment, $H = 0.15 \; m$, the Rayleigh number of the problem is

$$Ra = \frac{g\,\beta\,\Delta T\,H^3}{(\mu/\rho)(\kappa/\rho c)} = 1.89 \times 10^8. \qquad (7.53)$$

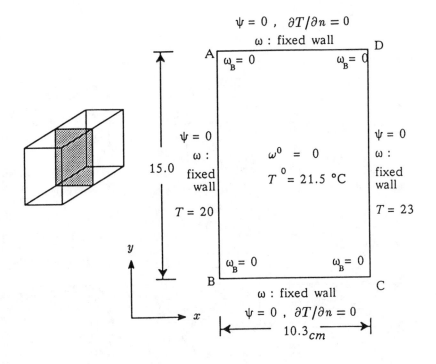

Figure 7.2: Cross section of an experimental vessel.

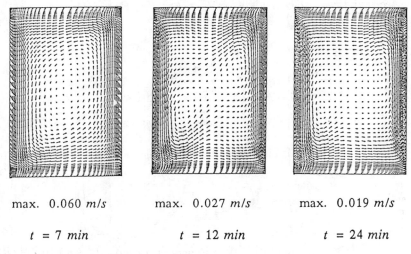

max. 0.060 *m/s*   max. 0.027 *m/s*   max. 0.019 *m/s*

t = 7 *min*   *t* = 12 *min*   *t* = 24 *min*

Figure 7.3: Calculated velocity vectors ($Ra = 1.89 \times 10^8$).

Calculated results are summarized in Figures 7.3 and 7.4. The thermal fluid motion is induced and evolves with the passage of time.

$t = 7\ min$

$t = 12\ min$

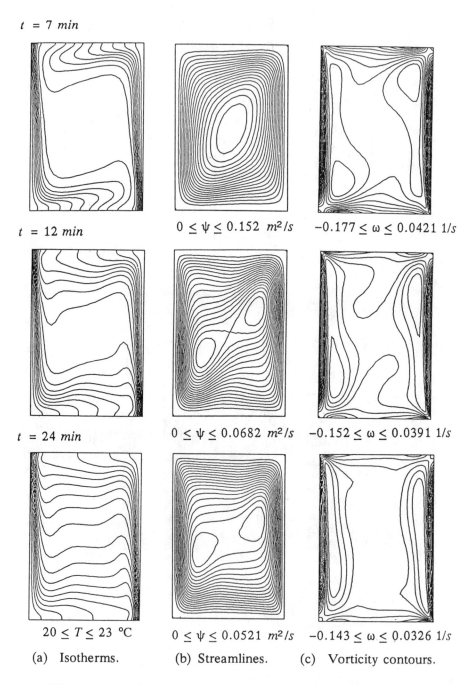

$0 \leq \psi \leq 0.152\ m^2/s$ $-0.177 \leq \omega \leq 0.0421\ 1/s$

$t = 24\ min$

$0 \leq \psi \leq 0.0682\ m^2/s$ $-0.152 \leq \omega \leq 0.0391\ 1/s$

$20 \leq T \leq 23\ °C$ $0 \leq \psi \leq 0.0521\ m^2/s$ $-0.143 \leq \omega \leq 0.0326\ 1/s$

(a)   Isotherms.   (b)   Streamlines.   (c)   Vorticity contours.

Figure 7.4:   Calculated evolution of natural convection in the compartment ($Ra = 1.89 \times 10^8$).

7.4.2   Bénard cell

We consider natural convection of water in a closed shallow vessel, as shown in Figure 7.5. The vessel is heated from below, so that the temperature difference between top and bottom is 1 °C. With $H = 0.01\ m$ and $\Delta T = 1\ °C$, the corresponding Rayleigh number is $Ra = 20250$.

Calculated results are shown in Figure 7.6. We can see the twelve Bénard cells generated in the vessel.

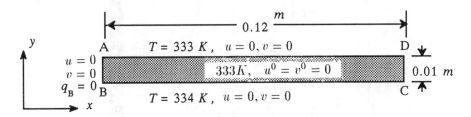

(a)   Boundary conditions.

(b)   Finite element mesh.

Figure 7.5: Cross section of a shallow vessel.

(a)   Isotherms ($333 \leq T \leq 334\ K$).

(b)   Streamlines ($-4.47 \times 10^{-4} \leq \psi \leq 4.32 \times 10^{-4}\ m^2/s$).

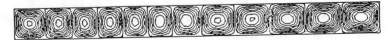

(c)   Vorticity contours ($-0.663 \leq \omega \leq 0.674\ 1/s$).

Figure 7.6: Calculated Bénard convection in quasi-steady state ($Ra = 20250$).

7.4.3   Forced thermal convection

We consider vertical forced convection of water between two parallel plates at distance 0.01 m apart and of 0.3 m long, as shown in Figure 7.7. The plates

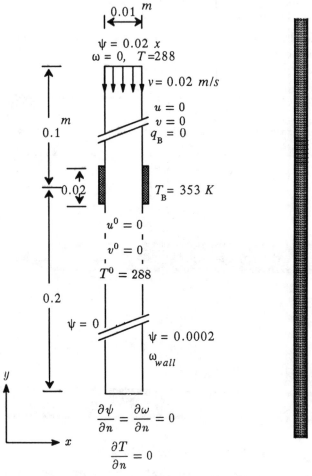

(a) Boundary conditions.   (b) Finite element mesh.

Figure 7.7: Vertical channel between two parallel plates.

are assumed to be adiabatic. Water at a temperature of 288 $K(= 15\ °C)$ is poured into the channel from the top inlet at a velocity of 0.02 m/s. Away from the inlet, there are heating stations of 0.02 m in length, at a temperature of 353 $K(= 80\ °C)$ on both sides.

Physical constants used are; $\nu = 1.14 \times 10^{-6}\ m^2/s$, $\lambda = 1.40 \times 10^{-7}\ m^2/s$, $\beta = 1.5 \times 10^{-4}\ 1/K$ and $g = 9.81\ m/s^2$. With $L = 0.01\ m$, $U = 0.02\ m/s$ and $\Delta T = 65\ °C$, the corresponding Reynolds and Froude numbers are $Re = 175$ and $Fr = 0.418$. The time increment in the computation is $\Delta t = 0.1\ s$.

Calculated results are presented in Figure 7.8.

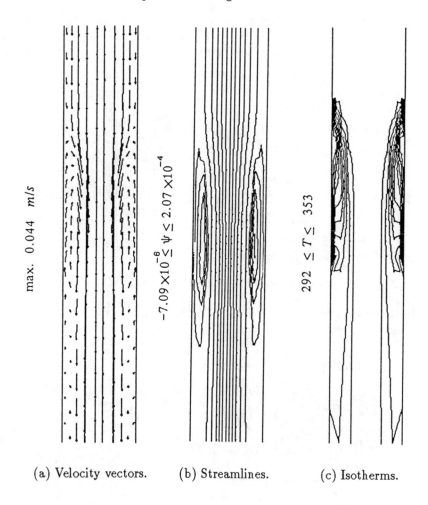

(a) Velocity vectors.   (b) Streamlines.   (c) Isotherms.

Figure 7.8: Calculated forced thermal convection near the heating spot
($t = 20\ s$).

Exercises

7.1 Using the computer program, solve the problem of natural convection shown in Figure 7.9 for two cases; (i) $Ra = 10^4$ and (ii) 10^5, with the Prandtl number $Pr = 1.0$, following the instructions below. Use the mesh

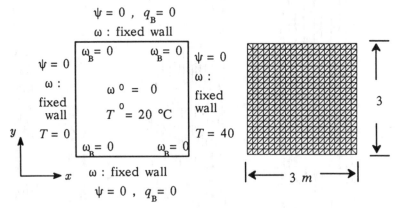

$$\psi = 0 , \quad q_B = 0$$
$$\omega : \text{fixed wall}$$

$\omega_B = 0 \qquad \omega_B = 0$

$\psi = 0$

$\omega :$

fixed wall

$\omega^0 = 0$

$T^0 = 20\ °C$

$\psi = 0$

$\omega :$

fixed wall

$T = 40$

y $\quad T = 0$

$\omega_B = 0 \qquad \omega_B = 0$

x $\quad \omega : \text{fixed wall}$

$$\psi = 0 , \quad q_B = 0$$

3

$\leftarrow\!\!\longrightarrow$ 3 m $\longrightarrow$

Figure 7.9: Problem of natural convection.

in the figure and take $\Delta t = 0.5$ s. Compare your results with the calculated results presented in Figure 7.10. Draw profiles of the streamfunction

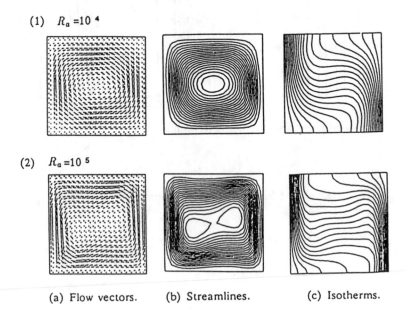

(1) $R_a = 10^4$

(2) $R_a = 10^5$

(a) Flow vectors.   (b) Streamlines.   (c) Isotherms.

Figure 7.10: Calculated results ($Pr = 1.0$).

and temperature along the horizontal or vertical center lines, as shown in Figure 7.11.

(i) $Ra = 10^4$. Put the gravitational acceleration $g = 9.8\ m/s^2$, the volumetric expansion coefficient $\beta = 0.21 \times 10^{-3}\ 1/K$, the temperature difference $\Delta T = 40\ K$, the representative length $L = 3\ m$. From the relation

$$Ra = \frac{g\,\beta\,\Delta T\,L^3}{\nu\,\lambda},$$

we know $\nu\lambda = 2.22 \times 10^{-4}\ m^4/s^2$. From the relation $Pr = \nu/\lambda = 1.0$, we have $\nu = \lambda = 1.49 \times 10^{-2}\ m^2/s$.

(ii) $Ra = 10^5$. We know $\nu\lambda = 2.22 \times 10^{-5}\ m^4/s^2$ and $\nu = \lambda = 4.71 \times 10^{-3}\ m^2/s$.

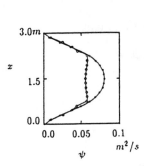

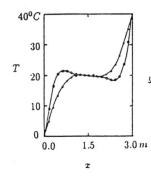

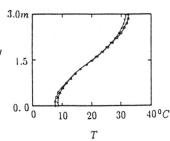

(a) Horizontal distribution of ψ along $y = 1.5$

(b) Distribution of T along $y = 1.5$

(c) Distribution of T along $x = 1.5$

Figure 7.11: Calculated profiles of streamfunction and temperature (black circles $Ra = 10^4$, white circles $Ra = 10^5$).

7.2 When the Rayleigh number is small, thermal energy is transported principally by conduction. As the number increases, the convection develops and thermal energy is transported simultaneously by the conduction and convection. The transition of its transport mechanism occurs at the critical Rayleigh number $Ra_{cri} = 1700$, as illustrated in Figure 1.7.

Solve the problem of natural heat transfer in a cup-shaped vessel, as shown in Figure 7.12, for the three cases; $Ra = 10^3, 10^4$ and 10^5, with $Pr = 1$ for all the cases. The fluid is heated from below at the temperature $60\ ^\circ C$ and the surface is exposed to the air at the temperature $15\ ^\circ C$. The free surface of the fluid in the vessel is cooled according to the Newton's law with the heat transfer coefficient $h = 0.04\ J/m^2 \cdot s \cdot K$. Take $g = 9.8\ m/s^2, \beta = 0.21 \times 10^{-3}\ 1/K, \Delta T = 45\ K$, and $L = 3\ m$. Put $\Delta t = 0.2\ s$.

Compare your results with the calculated results presented in Figure 7.13.

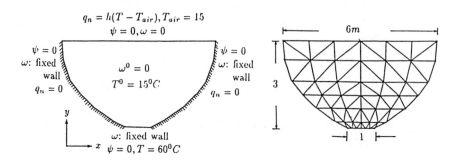

(a)   Problem statement.   (b)   Finite element mesh.

Figure 7.12: Problem of natural convection.

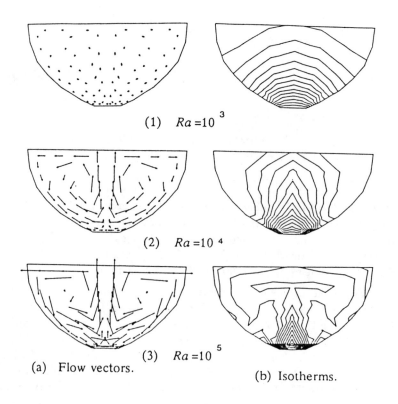

(1) $Ra = 10^3$

(2) $Ra = 10^4$

(3) $Ra = 10^5$

(a)   Flow vectors.   (b) Isotherms.

Figure 7.13: Calculated results ($Pr = 1.0$).

Chapter 8

MASS TRANSPORT

This chapter deals with the finite element analysis of isothermal mass transfer in a viscous fluid. A soluble mass diffuses into the liquid. When the liquid is in motion, the mass is transported simultaneously by diffusion and convection. The natural convective motion of the fluid is induced by the buoyancy due to the non-uniform density variations of the fluid under the action of gravity. It is assumed that the density depends only on the concentration of the mass under consideration. The equation of mass diffusion is coupled to the equations of viscous fluid flow. As numerical examples, we shall present a discharge of smoke from a chimney into the atmosphere, and density-dependent convective diffusion of two mixing viscous fluids.

8.1 Governing Equations

We shall consider a solution of mass in a fluid. We shall assume that the density, viscosity, diffusion coefficient are independent of the concentration except a term which describes the buoyancy F_y per a unit mass under the action of gravity in the equations of motion:

$$\frac{\partial u}{\partial t} + u \frac{\partial u}{\partial x} + v \frac{\partial u}{\partial y} = -\frac{1}{\rho} \frac{\partial p}{\partial x} + \nu \nabla^2 u \,, \qquad (8.1)$$

$$\frac{\partial v}{\partial t} + u \frac{\partial v}{\partial x} + v \frac{\partial v}{\partial y} = -\frac{1}{\rho} \frac{\partial p}{\partial y} + \nu \nabla^2 v + F_y(C) \,. \qquad (8.2)$$

We shall derive the expression for F_y, which depends on the concentration $C\,(kg/m^3)$. To this end, we consider a fluid at concentration C_0 and its density ρ_0 in a container, as shown in Figure 8.1 The ordinate y is taken in the opposite direction to the gravitational acceleration g. We assume that a lump of the fluid in the container changes its concentration into C, with density $\rho\,(kg/m^3)$. Then, the buoyancy force of magnitude $(\rho_0 - \rho)\,g\,(N/m^3)$ is exerted on the lump of the fluid. If the density difference is small enough, the equation of state of the fluid can be expressed as follows.

$$\frac{\rho}{\rho_0} = 1 + \sigma(C - C_0), \tag{8.3}$$

with a proportional constant σ (m^3/kg). The buoyancy per unit mass is therefore given by

$$F_y = (\rho_0 - \rho)g/\rho_0 = \sigma g(C - C_0). \tag{8.4}$$

The continuity equation remains unchanged as given by (6.3) under the above assumption, namely,

$$\frac{\partial u}{\partial x} + \frac{\partial v}{\partial y} = 0. \tag{8.5}$$

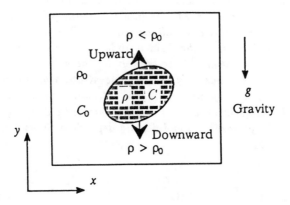

Figure 8.1: Buoyancy induced by non-uniform concentration.

With expression (8.4), the basic equations describing the fluid flow are written in terms of the streamfunction and vorticity as follows.

$$\nabla^2 \psi = -\omega, \tag{8.6}$$

$$\frac{\partial \omega}{\partial t} + \frac{\partial \psi}{\partial y}\frac{\partial \omega}{\partial x} - \frac{\partial \psi}{\partial x}\frac{\partial \omega}{\partial y} = \nu \nabla^2 \omega - g\sigma \frac{\partial C}{\partial x}. \tag{8.7}$$

Boundary conditions concerning ψ and ω are given as in Figure 6.1 and are as discussed in Section 6.3.

The solute is transported simultaneously by diffusion and convection. The diffusion is assumed to obey the *Fick's law* with the *diffusion coefficient* denoted by η (m^2/s). Components of the *total mass flux* $(kg/m^3 \cdot s)$, as a sum of the diffusive and convective fluxes, are given by

$$J_x = -\eta \frac{\partial C}{\partial x} + uC, \tag{8.8}$$

$$J_y = -\eta \frac{\partial C}{\partial y} + vC. \tag{8.9}$$

The law of conservation of mass is described by the equation

$$\frac{\partial C}{\partial t} + \frac{\partial J_x}{\partial x} + \frac{\partial J_y}{\partial y} = S , \qquad (8.10)$$

where S is the intensity of the mass source $(kg/m^3 \cdot s)$. Substituting (8.8) and (8.9) into this equation, we obtain the following *convective diffusion equation*:

$$\frac{\partial C}{\partial t} + \frac{\partial \psi}{\partial y}\frac{\partial C}{\partial x} - \frac{\partial \psi}{\partial x}\frac{\partial C}{\partial y} = \eta \nabla^2 C + S . \qquad (8.11)$$

Boundary conditions concerning C are given as follows: The concentration is specified on part of the boundary, and the diffussive flux in the exterior normal direction is specified on the rest of the boundary, namely,

$$C = C_B \qquad \text{on} \qquad \Gamma_C , \qquad (8.12)$$

$$-\eta \frac{\partial C}{\partial n} = j_B \qquad \text{on} \qquad \Gamma_j . \qquad (8.13)$$

If the boundary is a *non-absorbing* or *reflexive boundary*, then we can put $j_B = 0$. If a discharge of the mass is instantaneous, the amount of the discharge can be considered in the initial condition $C = C^0$.

8.2   Finite Element Discretisation

We shall apply the conventional Galerkin finite element method to discretising of the streamfunction equation (8.6), the vorticity transport equation (8.7), and the convective diffusion equation (8.11). To this end, let $\delta\psi, \delta\omega$ and δC be weighting functions, being arbitrary expect when $\delta\psi = 0$ on Γ_ψ, $\delta\omega = 0$ on $\Gamma_\omega \cup \Gamma_w$, and $\delta C = 0$ on Γ_C. We start the finite element formulation with the weighted residual forms of these equations:

$$\int_\Omega \delta\psi \, \nabla^2 \psi \, d\Omega + \int_\Omega \delta\psi \, \omega \, d\Omega = 0 , \qquad (8.14)$$

$$\int_\Omega \delta\omega \frac{\partial \omega}{\partial t} \, d\Omega + \int_\Omega \delta\omega \left(\frac{\partial \psi}{\partial y}\frac{\partial \omega}{\partial x} - \frac{\partial \psi}{\partial x}\frac{\partial \omega}{\partial y} \right) d\Omega$$

$$+ \int_\Omega \delta\omega \, g \, \sigma \frac{\partial C}{\partial x} \, d\Omega - \int_\Omega \delta\omega \, \nu \, \nabla^2 \omega \, d\Omega = 0 , \qquad (8.15)$$

$$\int_\Omega \delta C \frac{\partial C}{\partial t} \, d\Omega + \int_\Omega \delta C \left(\frac{\partial \psi}{\partial y}\frac{\partial C}{\partial x} - \frac{\partial \psi}{\partial x}\frac{\partial C}{\partial y} \right) d\Omega$$

$$- \int_\Omega \delta C \, \eta \, \nabla^2 C \, d\Omega - \int_\Omega \delta C \, S \, d\Omega = 0 . \qquad (8.16)$$

Integration by parts of the terms involving the Laplacian yields

$$\int_\Omega \nabla\delta\psi \cdot \nabla\psi \, d\Omega - \int_\Omega \delta\psi \, \omega \, d\Omega - \int_{\Gamma_s} \delta\psi \frac{\partial \psi}{\partial n} \, d\Gamma = 0 , \qquad (8.17)$$

$$\int_\Omega \delta w \frac{\partial w}{\partial t}\, d\Omega + \int_\Omega \delta w \left(\frac{\partial \psi}{\partial y}\frac{\partial w}{\partial x} - \frac{\partial \psi}{\partial x}\frac{\partial w}{\partial y} \right) d\Omega + \int_\Omega \nu\, \nabla \delta w \cdot \nabla w\, d\Omega$$

$$+ \int_\Omega \delta w\, g\, \sigma \frac{\partial C}{\partial x}\, d\Omega - \int_{\Gamma_x} \nu\, \delta w \frac{\partial w}{\partial n}\, d\Gamma = 0\,, \tag{8.18}$$

$$\int_\Omega \delta C \frac{\partial C}{\partial t}\, d\Omega + \int_\Omega \delta C \left(\frac{\partial \psi}{\partial y}\frac{\partial C}{\partial x} - \frac{\partial \psi}{\partial x}\frac{\partial C}{\partial y} \right) d\Omega + \int_\Omega \eta\, \nabla \delta C \cdot \nabla C\, d\Omega$$

$$- \int_\Omega \delta C\, S\, d\Omega - \int_{\Gamma_j} j_B\, \delta C\, d\Gamma = 0\,. \tag{8.19}$$

The domain Ω is divided into triangular finite elements. Inside each triangle e having its three vertices as the element nodes with their local node numbers 1, 2, 3; the unknown streamfunction, vorticity and concentration are linearly interpolated as follows.

$$\psi = \sum_\alpha \phi_\alpha\, \psi_\alpha\,, \qquad \delta\psi = \sum_\alpha \phi_\alpha\, \delta\psi_\alpha\,,$$

$$w = \sum_\alpha \phi_\alpha\, w_\alpha\,, \qquad \delta w = \sum_\alpha \phi_\alpha\, \delta w_\alpha\,, \tag{8.20}$$

$$C = \sum_\alpha \phi_\alpha\, C_\alpha\,, \qquad \delta C = \sum_\alpha \phi_\alpha\, \delta C_\alpha\,,$$

where the summation notation $\sum_\alpha$ indicates summation for all α ($\alpha = 1, 2, 3$), and ϕ_α are linear interpolation functions given by

$$\phi_\alpha = \frac{1}{2\,\Delta^e}\,(a_\alpha + b_\alpha x + c_\alpha y)$$

with the area of the triangle Δ^e; $\psi_\alpha, w_\alpha, C_\alpha$ are nodal values of the corresponding unknowns; and $\delta\psi_\alpha, \delta w_\alpha, \delta C_\alpha$ are their arbitrary nodal variations.

The interpolations (8.20) are substituted into (8.17)-(8.19). From the continuity in the interpolations of ψ, w, C and the arbitrariness of $\delta\psi_\alpha, \delta w_\alpha, \delta C_\alpha$, the element equations become

$$\sum_{\beta=1}^{3} D_{\alpha\beta}^e\, \psi_\beta - \sum_{\beta=1}^{3} M_{\alpha\beta}^e\, w_\beta - \Gamma_{s\alpha}^e = 0\,, \tag{8.21}$$

$$\sum_{\beta=1}^{3} M_{\alpha\beta}^e\, \dot{w}_\beta + \sum_{\beta=1}^{3} A_{\alpha\beta}^e\, w_\beta + \nu \sum_{\beta=1}^{3} D_{\alpha\beta}^e\, w_\beta$$

$$+ G_\alpha^e(C) - \Gamma_{\chi\alpha}^e = 0\,, \tag{8.22}$$

$$\sum_{\beta=1}^{3} M_{\alpha\beta}^e\, \dot{C}_\beta + \sum_{\beta=1}^{3} A_{\alpha\beta}^e\, C_\beta + \eta \sum_{\beta=1}^{3} D_{\alpha\beta}^e\, C_\beta$$

$$- S_\alpha^e + \Gamma_{j\alpha}^e = 0\,, \qquad (\alpha = 1, 2, 3) \tag{8.23}$$

where the coefficients are given by

$$M_{\alpha\beta}^e = \int_e \phi_\alpha\, \phi_\beta\, d\Omega = \frac{\Delta^e}{12}\,(1 + \delta_{\alpha\beta})\,, \tag{8.24}$$

$$D^e_{\alpha\beta} = \int_e \left(\frac{\partial \phi_\alpha}{\partial x}\frac{\partial \phi_\beta}{\partial x} + \frac{\partial \phi_\alpha}{\partial y}\frac{\partial \phi_\beta}{\partial y} \right) d\Omega = \frac{1}{4\Delta^e}(b_\alpha b_\beta + c_\alpha c_\beta) , \quad (8.25)$$

$$A^e_{\alpha\beta} = \int_e \phi_\alpha \left(\sum_{\gamma=1}^{3} \frac{\partial \phi_\gamma}{\partial y}\psi_\gamma \frac{\partial \phi_\beta}{\partial x} - \sum_{\gamma=1}^{3} \frac{\partial \phi_\gamma}{\partial x}\phi_\gamma \frac{\partial \phi_\beta}{\partial y} \right) d\Omega$$

$$= \frac{1}{12\Delta^e} \sum_{\gamma=1}^{3}(c_\gamma b_\beta - b_\gamma c_\beta)\psi_\gamma , \quad (8.26)$$

$$G^e_\alpha = g\beta \int_e \phi_\alpha \frac{\partial}{\partial x}\left(\sum_{\gamma=1}^{3} \phi_\gamma C_\gamma \right) d\Omega = \frac{g\beta}{6} \sum_{\gamma=1}^{3} b_\gamma C_\gamma , \quad (8.27)$$

$$S^e_\alpha = \int_e \phi_\alpha \left(\sum_{\gamma=1}^{3} \phi_\gamma S_\gamma \right) d\Omega = \frac{1}{12\Delta^e}\left(S_\alpha + \sum_{\gamma=1}^{3} S_\gamma \right), \quad (8.28)$$

$$\Gamma^e_{s\alpha} = \int_{\Gamma^e_s} \phi_\alpha \frac{\partial \psi}{\partial n} d\Gamma = -\frac{V_s}{2} \mid \Gamma^e_s \mid \begin{bmatrix} 0 \\ 1 \\ 1 \end{bmatrix}, \quad (8.29)$$

$$\Gamma^e_{\chi\alpha} = \int_{\Gamma^e_\chi} \nu\,\phi_\alpha \frac{\partial \omega}{\partial n} d\Gamma = \nu\frac{\chi_n}{2} \mid \Gamma^e_\chi \mid \begin{bmatrix} 0 \\ 1 \\ 1 \end{bmatrix}, \quad (8.30)$$

$$\Gamma^e_{j\alpha} = \int_{\Gamma^e_j} \phi_\alpha\, j_B\, d\Gamma = \frac{j_B}{2} \mid \Gamma^e_j \mid \begin{bmatrix} 0 \\ 1 \\ 1 \end{bmatrix}, \quad (8.31)$$

with $\Gamma^e_j = \Gamma_j \cup \partial e$ under a similar situation to the one presented in Figure 3.8.

After assembling all the element equations over the whole domain, we obtain the total equations in the form:

$$[D]\{\psi\} - [M]\{\omega\} - \{\Gamma_s\} = \{0\} , \quad (8.32)$$

$$[M]\{\dot{\omega}\} + [A(\psi)]\{\omega\} + \nu[D]\{\omega\}$$
$$+ \{G(C)\} - \{\Gamma_\chi\} = \{0\} , \quad (8.33)$$

$$[M]\{\dot{C}\} + [A(\psi)]\{C\} + \eta[D]\{C\}$$
$$- \{S\} + \{\Gamma_j\} = \{0\} , \quad (8.34)$$

where $[D]$, $[M]$, $[A]$ denote the total matrices, $\{G\}$ corresponds to the buoyancy term; $\{\psi\}$, $\{\omega\}$, $\{C\}$ are the nodal unknown column vectors. Other column vectors are due to the respective boundary conditions. One interprets (8.32)-(8.34) as a nonlinear system of first-order ordinary differential equations for unknown $\{\psi\}$, $\{\omega\}$ and $\{C\}$.

8.3 Computational Scheme

We shall consider the discretisation of the total equations with respect to time. To this end, we apply the semi-implicit scheme to the initial value problem. The time derivatives of the nodal vorticity and concentration are approximated by the following finite differences.

$$\dot{\omega}_\beta = \frac{d\omega_\beta}{dt} \approx \frac{\omega_\beta^{n+1} - \omega_\beta^n}{\Delta t}, \tag{8.35}$$

$$\dot{C}_\beta = \frac{dC_\beta}{dt} \approx \frac{C_\beta^{n+1} - C_\beta^n}{\Delta t}, \tag{8.36}$$

where ω_β^n and C_β^n denote the nodal vorticity and concentration, respectively, at the time level t_n, defined by $t_{n+1} = t_n + \Delta t$ $(n = 0, 1, 2, \ldots)$ with the time increment Δt.

We consider the approximation of (8.32)-(8.34) at the time level t_{n+1} by replacing the time derivatives with (8.35) and (8.36). This results in

$$[D]\{\psi^{n+1}\} = [M]\{\omega^n\} + \{\Gamma_s\}, \tag{8.37}$$

$$\frac{1}{\Delta t}[M]\{\omega^{n+1}\} + \nu[D]\{\omega^{n+1}\}$$

$$= \frac{1}{\Delta t}[M]\{\omega^n\} - [A(\psi^{n+1})]\{\omega^n\} - \{G(C^n)\} + \{\Gamma_\chi^{n+1}\}, \tag{8.38}$$

$$\frac{1}{\Delta t}[M]\{C^{n+1}\} + \eta[D]\{C^{n+1}\}$$

$$= \frac{1}{\Delta t}[M]\{C^n\} - [A(\psi^{n+1})]\{C^n\} + \{S^{n+1}\} - \{\Gamma_j^{n+1}\}. \tag{8.39}$$

The artificial kinematic viscosity as well as the artificial diffusivity are introduced in the computation. Instead of ν and η, we use

$$\nu_{ax} = \nu + \frac{1}{2}u^2\Delta t, \quad \nu_{ay} = \nu + \frac{1}{2}v^2\Delta t, \tag{8.40}$$

$$\eta_{ax} = \eta + \frac{1}{2}u^2\Delta t, \quad \eta_{ay} = \eta + \frac{1}{2}v^2\Delta t. \tag{8.41}$$

In terms of Algol-like statements, the computation proceeds as follows.

Read *topological data.*

Read *material constants and boundary conditions.*

Set initial values.

For $n = 0, 1, 2, \ldots$, **until** *satisfied, do:*

> *Calculate* $\{\psi^{n+1}\}$.
> *Insert the boundary values* ω_{wall} *on* Γ_w.
> *Calculate* $\{\omega^{n+1}\}$.
> *Calculate* $\{C^{n+1}\}$.

The algorithm is not complete until some stopping criteria are specified for the iteration counter n. When the flow duration is given, n goes to the integer $N = (t_f - t_0)/\Delta t$ with the final time t_f. Otherwise, the iteration may continue until the calculated density-dependent fluid motion fully develops.

We shall consider non-dimensional forms of the governing equations. For forced convection with

$$C^* = \frac{C - C^0}{\Delta C} ,$$ (8.42)

it follows from (8.6), (8.7) and (8.11) that

$$\nabla^{*2} \psi^* = -\omega^* ,$$ (8.43)

$$\frac{\partial \omega^*}{\partial t^*} + \frac{\partial \psi^*}{\partial y^*} \frac{\partial \omega^*}{\partial x^*} - \frac{\partial \psi^*}{\partial x^*} \frac{\partial \omega^*}{\partial y^*} = \frac{1}{Re} \nabla^{*2} \omega^* - \frac{1}{Fr} \frac{\partial C^*}{\partial x^*} ,$$ (8.44)

$$\frac{\partial C^*}{\partial t^*} + \frac{\partial \psi^*}{\partial y^*} \frac{\partial C^*}{\partial x^*} - \frac{\partial \psi^*}{\partial x^*} \frac{\partial C^*}{\partial y^*} = \frac{1}{Re\,Sc} \nabla^{*2} C^* ,$$ (8.45)

with the *Froude number* $Fr = U^2/(g\sigma\Delta CL)$ and the *Schmidt number* $Sc = \nu/\eta$. For *density-dependent* or *free convection*, we have

$$\nabla^{*2} \psi^* = -\omega^* ,$$ (8.46)

$$\frac{\partial \omega^*}{\partial t^*} + \frac{\partial \psi^*}{\partial y^*} \frac{\partial \omega^*}{\partial x^*} - \frac{\partial \psi^*}{\partial x^*} \frac{\partial \omega^*}{\partial y^*} = \nabla^{*2} \omega^* - Gr \frac{\partial C^*}{\partial x^*} ,$$ (8.47)

$$\frac{\partial C^*}{\partial t^*} + \frac{\partial \psi^*}{\partial y^*} \frac{\partial C^*}{\partial x^*} - \frac{\partial \psi^*}{\partial x^*} \frac{\partial C^*}{\partial y^*} = \frac{1}{Sc} \nabla^{*2} C^* ,$$ (8.48)

with the *Grashof number* $Gr = g\sigma\Delta CL^3/\nu^2$ for the mass transfer.

8.4   Numerical Examples

8.4.1   Smoke advection in the air

We consider the convective diffusion of discharged smoke from a chimney into the atmosphere. Let $C\,(kg/m^3)$ be the concentration of substances in the smoke. The chimney pot is located $7\,m$ high above the ground. The wind at the mean velocity $U = 2\,m/s$ is assumed to blow horizontally from the left to right, as shown in Figure 8.2. Downstream from the chimney is located a structural obstacle with cross section of 6 m high and 4 m wide. For the sake of simplicity, we assume that the flow does not depend on the concentration.

The intensity of the smoke discharge is $S = 0.001\,kg/m^3{\cdot}s$. Physical constants for the wind are; $\nu = 1.45 \times 10^{-5}\,m^2/s$, $\eta = 1.58 \times 10^{-5}\,m^2/s$, $g = 9.81\,m/s^2$ and $\sigma = 0$. With the representative height of the structure $H = 6\,m$, the Reynolds and Peclet numbers for the problem are as follows

$$Re = 828000 , \qquad Pe = \frac{\nu}{\eta} = 0.918 .$$ (8.49)

The boundary conditions are indicated in Figure 8.2. We impose the condition that the concentration contours perpendicularly intersect the efflux boundary CD on the right-hand side. We assume that the atmosphere is initially motionless. We used the time increment $\Delta t = 0.1\,s$. The finite element mesh is presented in Figure 8.3.

The calculated results are summarized in Figure 8.4. As the wind develops, the smoke is transported mainly by convection.

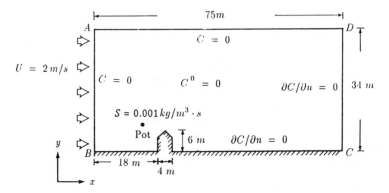

Figure 8.2: Discharge of smoke into the atmosphere.

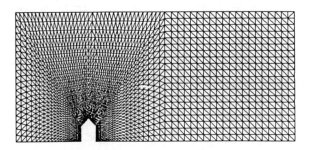

Figure 8.3: Finite element mesh.

8.4.2   Density-dependent viscous flow

We consider a closed experimental vessel with rectangular cross section, as shown
in Figure 8.5. The vessel is divided into two cells by an impervious thin plate. In
the right cell is contained pure water at the initial concentration $C_0 = 0 \ kg/m^3$,
while in the left cell is contained a solution at the unit initial concentration
$C_0 = 1 \ kg/m^3$. We shall calculate the mixing process of the solution after the
intermediate plate is removed.

We assume that $\nu = 1.14 \times 10^{-6} \ m^2/s$, $\eta = 0.90 \times 10^{-9} \ m^2/s$ and $\sigma =$
$0.05 \ m^3/kg$. With the representative length $L = 0.02 \ m$ and the concentration
difference $\Delta C = 1 \ kg/m^3$, the Rayleigh number is

$$Ra = \frac{g \sigma \Delta C \, L^3}{\nu \eta} = 1.03 \times 10^{11} \ . \tag{8.50}$$

We take $\Delta t = 0.01 \ s$. The calculated results are shown in Figure 8.6.

(a) Concentration. $0 \leq C \leq 74.5$

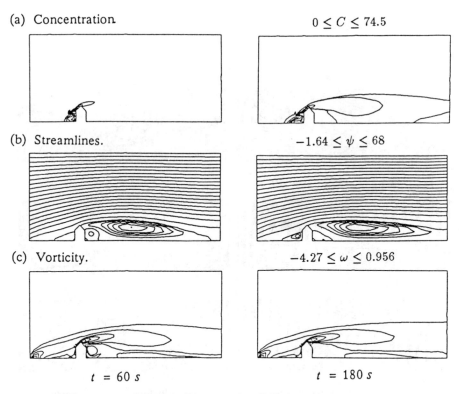

(b) Streamlines. $-1.64 \leq \psi \leq 68$

(c) Vorticity. $-4.27 \leq \omega \leq 0.956$

$t = 60\ s$ $t = 180\ s$

Figure 8.4: Calculated convective diffusion of smoke
($Re = 828000$, $Pe = 0.918$).

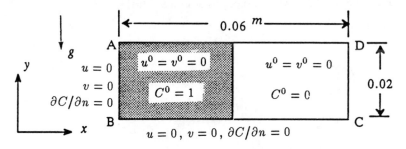

(a)   Boundary conditions.

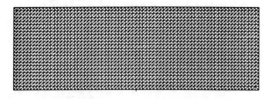

(b)   Finite element mesh.

Figure 8.5: Twin-cell vessel.

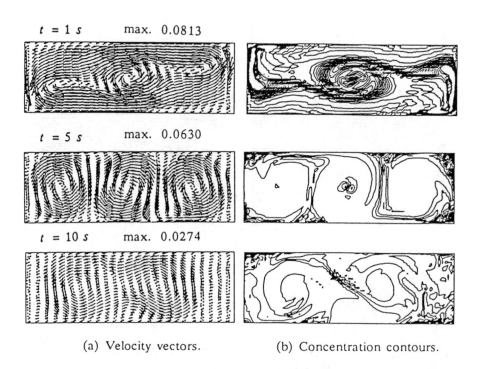

(a) Velocity vectors.   (b) Concentration contours.

Figure 8.6: Calculated mixing in aquious solution ($Ra = 1.03 \times 10^{11}$).

Exercises

8.1 Derive the non-dimensional forms (8.43) - (8.48).

8.2 Using the attached computer program, solve the problem of a discharge of
smoke, as shown in Figure 8.2. Use the coarse mesh presented in Figure
8.7, and take $\nu = 0.001$ m^2/s and $\Delta t = 0.1$ s. Compare your results with
the calculated results presented in Figure 8.8.

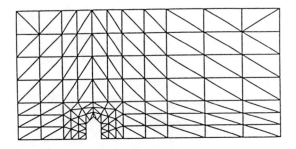

Figure 8.7: Coarse finite element mesh.

(a) Flow vectors.

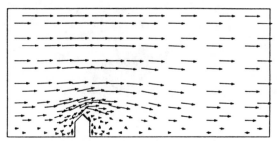

(b) Concentration contours.

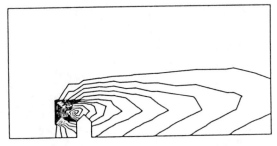

Figure 8.8: Calculated results ($t = 30$ s)

Chapter 9

TIDAL CURRENT

The tidal current is modelled in two dimensions. The shallow water equations are approximated by the Galerkin finite element method with linear triangular elements. Two-step selectively lumped explicit scheme, so-called the Kawahara scheme, is used for the discretisation of time. The scheme is applied to the simulation of tidal currents in channels and bays.

9.1 Governing Equations

Horizontal flow of the seawater induced by the motion of the moon and the sun is called the *tidal current*. The vertical motion of the sea surface is called the *tide*. The tide has about twelve hours and twenty five minutes of period at the place, where the tidal motion is semi-diurnal. The continual motion of the sea water in some definite direction is called the *ocean current*. In this section, we shall summarize the basic equations which describe the tidal current.

We shall consider an area of the sea with the rectangular coordinates x, y, z (m) as illustrated in Figure 9.1. Let the surface of the mean sea level designate the reference plane $z = 0$. Let $h(x, y)$ be the depth measured from the mean sea level to the sea bed. The elevation from the mean sea level to the temporary sea surface is called the *tidal elevation*, and it is denoted by $\zeta(x, y, t)$ with the time variable t (s). The total depth of the sea is therefore given by $H = h + \zeta$.

We assume that the density of the seawater ρ (kg/m^3) is constant. With the velocity components u, v (m/s) of the current to the east and north respectively, the components U, V (m/s) of the vertically averaged velocity are expressed by

$$U = \frac{1}{H} \int_{-h}^{\zeta} u \, dz , \qquad V = \frac{1}{H} \int_{-h}^{\zeta} v \, dz . \qquad (9.1)$$

The equation of continuity can then be written as follows.

$$\frac{\partial \zeta}{\partial t} + \frac{\partial (HU)}{\partial x} + \frac{\partial (HV)}{\partial y} = 0 . \qquad (9.2)$$

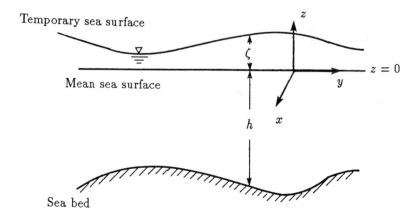

Figure 9.1: Vertical cross section of the sea.

Suppose that, as regards the motion of the seawater, the acceleration in its vertical direction is negligibly small. Based on the assumption that the pressure is hydrostatic, the equations of motion can be given approximately as follows, see *e.g.* Pinder and Gray [1977].

$$\frac{\partial U}{\partial t} + U \frac{\partial U}{\partial x} + V \frac{\partial U}{\partial y} - fV + g \frac{\partial \zeta}{\partial x}$$

$$= \frac{1}{\rho H} \{ \frac{\partial}{\partial x} (\mu_e H \frac{\partial U}{\partial x}) + \frac{\partial}{\partial y} (\mu_e H \frac{\partial U}{\partial y}) \}$$

$$+ \frac{K W^2}{H} \cos \psi - \frac{gU(U^2 + V^2)^{1/2}}{H C^2}, \qquad (9.3)$$

$$\frac{\partial V}{\partial t} + U \frac{\partial V}{\partial x} + V \frac{\partial V}{\partial y} + fU + g \frac{\partial \zeta}{\partial y}$$

$$= \frac{1}{\rho H} \{ \frac{\partial}{\partial x} (\mu_e H \frac{\partial V}{\partial x}) + \frac{\partial}{\partial y} (\mu_e H \frac{\partial V}{\partial y}) \}$$

$$+ \frac{K W^2}{H} \sin \psi - \frac{gV(U^2 + V^2)^{1/2}}{H C^2}, \qquad (9.4)$$

where f is the Coriolis factor, given by $f = 2w \sin \phi$ with the angular velocity of the terrestrial rotation $w = 7.292 \times 10^{-5}$ (rad/s) and the latitude ϕ (rad); g is the gravity acceleration $= 9.81$ (m/s^2), μ_e is the eddy viscosity, K is the non-dimensional coefficient of a superficial force due to the wind blowing on the surface, W is the wind speed (m/s) 10 meters high above the sea surface, ψ is the angle of the wind direction from the east, and C is the Chezy coefficient $(m^{1/2}/s)$ of the friction on the sea bed. The coefficients are often given by

$$K = \begin{cases} 1.0 \times 10^{-3} & (W \le 5) \\ 1.5 \times 10^{-3} & (5 < W \le 15) \\ 2.0 \times 10^{-3} & (15 < W \le 20), \end{cases}$$

$$C = \frac{1}{n} h^{1/6} \tag{9.5}$$

with the Manning's coefficient of roughness n.

When the shearing stresses, the surface wind and frictional forces are neglected, the equations of motion become

$$\frac{\partial U}{\partial t} + U \frac{\partial U}{\partial x} + V \frac{\partial U}{\partial y} - fV + g \frac{\partial \zeta}{\partial x} = 0 , \tag{9.6}$$

$$\frac{\partial V}{\partial t} + U \frac{\partial V}{\partial x} + V \frac{\partial V}{\partial y} + fU + g \frac{\partial \zeta}{\partial y} = 0 . \tag{9.7}$$

These equations are called the *shallow water equations*. They are approximate equations and are usually applicable to coastal seas and estuaries, where the water depth is about less than $1/20$ of the wave length.

Two kinds of boundary conditions are considered: Along the seashore denoted by Γ_V, the normal component of the current velocity is equal to zero:

$$V_n = U n_x + V n_y = 0 \quad \text{on} \quad \Gamma_V , \tag{9.8}$$

where n_x, n_y are the components of the external unit normal to the boundary. Along the boundary Γ_ζ, where the region is open to the outer sea, the tidal elevation is prescribed. For the sake of simplicity we consider

$$\zeta(t) = A \sin\left(\frac{2\pi}{T} t\right) \quad \text{on} \quad \Gamma_\zeta , \tag{9.9}$$

where A is the amplitude of the monochromatic tidal wave (m), and $T = 12 + 5/12$ $(hours)$ being its period.

The initial condition in a real sea is difficult to know. Usually, the calculation starts with the calm sea: We suppose that

$$\zeta = 0 , \quad U = V = 0 \quad \text{at} \quad t = 0 , \quad \text{and in} \quad \Omega , \tag{9.10}$$

where Ω is the domain to be analysed. This trick is called a *cold start*.

9.2   Finite Element Discretisation

We shall apply the conventional Galerkin finite element method to the continuity equation (9.2) and the shallow water equations (9.6), (9.7) for the discretisation. To this end, let $\delta\zeta$ be weighting functions, being arbitrary but equal to zero on the boundary Γ_ζ. We start the finite element formulation with the weighted residual form of the continuity equation:

$$\int_\Omega \delta\zeta \frac{\partial \zeta}{\partial t} d\Omega + \int_\Omega \delta\zeta \left(\frac{\partial(HU)}{\partial x} + \frac{\partial(HV)}{\partial y} \right) d\Omega = 0 , \tag{9.11}$$

with $d\Omega = dxdy$. Integration by parts of the second term yields that

$$\int_\Omega \delta\zeta \frac{\partial \zeta}{\partial t} d\Omega - \int_\Omega H \left(\frac{\partial \delta\zeta}{\partial x} U + \frac{\partial \delta\zeta}{\partial y} V \right) d\Omega = - \int_{\Gamma_V} \delta\zeta H V_n d\Gamma \tag{9.12}$$

with the infinitesimal boundary length $d\Gamma$.

Let δU and δV be arbitrary weighting functions. The weighted residual forms corresponding to the equations of motion can be written as follows.

$$\int_\Omega \delta U \frac{\partial U}{\partial t}\, d\Omega + \int_\Omega \delta U\,(U\frac{\partial U}{\partial x} + V\frac{\partial U}{\partial y})\, d\Omega$$

$$- \int_\Omega \delta U\, f\, V\, d\Omega + g\int_\Omega \delta U\,\frac{\partial \zeta}{\partial x}\, d\Omega\ =\ 0\,, \tag{9.13}$$

$$\int_\Omega \delta V \frac{\partial V}{\partial t}\, d\Omega + \int_\Omega \delta V\,(U\frac{\partial U}{\partial x} + V\frac{\partial V}{\partial y})\, d\Omega$$

$$+ \int_\Omega \delta V\, f\, U\, d\Omega + g\int_\Omega \delta V\,\frac{\partial \zeta}{\partial y}\, d\Omega\ =\ 0\,. \tag{9.14}$$

The domain Ω is divided into triangular finite elements. Inside each triangle e with its three vertices as the element nodes 1, 2, 3; the tidal elevation as well as the mean velocity components are linearly interpolated as follows.

$$\zeta = \sum_\alpha \phi_\alpha \zeta_\alpha\,, \qquad \delta\zeta = \sum_\alpha \phi_\alpha \delta\zeta_\alpha\,,$$

$$U = \sum_\alpha \phi_\alpha U_\alpha\,, \qquad \delta U = \sum_\alpha \phi_\alpha \delta U_\alpha\,, \tag{9.15}$$

$$V = \sum_\alpha \phi_\alpha V_\alpha\,, \qquad \delta V = \sum_\alpha \phi_\alpha \delta V_\alpha\,,$$

where ϕ_α ($\alpha = 1,2,3$) are linear interpolation functions, given by

$$\phi_\alpha = \frac{1}{2\,\Delta^e}(a_\alpha + b_\alpha x + c_\alpha y)$$

with the area Δ^e of the triangle, and $\zeta_\alpha, U_\alpha, V_\alpha$ are nodal values of the corresponding unknowns.

The interpolations (9.15) are substituted into (9.12)-(9.14). From the continuity in the interpolations of ζ, U, V and the arbitrariness of $\delta\zeta_\alpha, \delta U_\alpha, \delta V_\alpha$, the element equations in each triangle become

$$\sum_\beta M^e_{\alpha\beta}\dot\zeta_\beta - \sum_{\beta,\gamma} X^e_{\beta\gamma\alpha} U_\beta H_\gamma - \sum_{\beta,\gamma} Y^e_{\beta\gamma\alpha} V_\beta H_\gamma$$

$$= -\sum_\beta R^e_{\alpha\beta} H_\beta\,, \tag{9.16}$$

$$\sum_\beta M^e_{\alpha\beta}\dot U_\beta + \sum_{\beta,\gamma} X^e_{\alpha\beta\gamma} U_\beta U_\gamma + \sum_{\beta,\gamma} Y^e_{\alpha\beta\gamma} V_\beta U_\gamma$$

$$- \sum_{\beta,\gamma} N^e_{\alpha\beta\gamma} f_\beta V_\gamma + g\sum_\beta P^e_{\alpha\beta}\zeta_\beta\ =\ 0\,, \tag{9.17}$$

$$\sum_\beta M^e_{\alpha\beta}\dot V_\beta + \sum_{\beta,\gamma} X^e_{\alpha\beta\gamma} U_\beta V_\gamma + \sum_{\beta,\gamma} Y^e_{\alpha\beta\gamma} V_\beta V_\gamma$$

$$- \sum_{\beta,\gamma} N^e_{\alpha\beta\gamma} f_\beta U_\gamma + g\sum_\beta Q^e_{\alpha\beta}\zeta_\beta\ =\ 0\,. \tag{9.18}$$

Here, the coefficients are given by

$$M^e_{\alpha\beta} = \int_e \phi_\alpha \phi_\beta \, d\Omega = \frac{\Delta^e}{12}(1 + \delta_{\alpha\beta}),$$

$$X^e_{\alpha\beta\gamma} = \int_e \phi_\alpha \phi_\beta \frac{\partial \phi_\gamma}{\partial x} \, d\Omega = \frac{b_\gamma}{24}(1 + \delta_{\alpha\beta}),$$

$$Y^e_{\alpha\beta\gamma} = \int_e \phi_\alpha \phi_\beta \frac{\partial \phi_\gamma}{\partial y} \, d\Omega = \frac{c_\gamma}{24}(1 + \delta_{\alpha\beta}),$$

$$N^e_{\alpha\beta\gamma} = \int_e \phi_\alpha \phi_\beta \phi_\gamma \, d\Omega$$

$$(\gamma)$$

$$= \frac{\Delta^e}{60} \, (\beta) \begin{vmatrix} 6 & 2 & 2 \\ 2 & 2 & 1 \\ 2 & 1 & 2 \end{vmatrix} , \begin{vmatrix} 2 & 2 & 1 \\ 2 & 6 & 2 \\ 1 & 2 & 2 \end{vmatrix} , \begin{vmatrix} 2 & 1 & 2 \\ 1 & 2 & 2 \\ 2 & 2 & 6 \end{vmatrix} ,$$

$$(\alpha = 1) \qquad (\alpha = 2) \qquad (\alpha = 3)$$

$$P^e_{\alpha\beta} = \int_e \phi_\alpha \frac{\partial \phi_\beta}{\partial x} \, d\Omega = \frac{b_\beta}{6},$$

$$Q^e_{\alpha\beta} = \int_e \phi_\alpha \frac{\partial \phi_\beta}{\partial x} \, d\Omega = \frac{c_\beta}{6},$$

$$R^e_{\alpha\beta} = \int_{\Gamma^e_V} \phi_\alpha \phi_\beta V_n \, d\Gamma = \frac{\Gamma^e_V}{6}(1 + \delta_{\alpha\beta}) V_n,$$

where $\Gamma^e_V = \Gamma_V \cap \partial e$.

After assembling all the element equations over the whole domain, we can obtain the total equations in the form:

$$[M]\{\dot{\zeta}\} - [H(U,V)]\{H\} = -[R]\{H\}, \qquad (9.19)$$
$$[M]\{\dot{U}\} + [A(U,V)]\{U\} - [N]\{V\} + g[P]\{\zeta\} = \{0\}, \quad (9.20)$$
$$[M]\{\dot{V}\} + [A(U,V)]\{V\} + [N]\{U\} + g[Q]\{\zeta\} = \{0\}. \quad (9.21)$$

One interprets them as a nonlinear system of first-order ordinary differential equations for unknown $\{H\} (= \{h\} + \{\zeta\})$, $\{U\}$ and $\{V\}$.

9.3   Computational Scheme

We shall consider discretisation of the total equations with respect to time. To this end, we present the finite difference method for the initial value problem of the differential equation

$$\frac{dy}{dx} = f(x,y) \quad \text{with} \quad y(x_0) = y^0,$$

with a sufficiently smooth function $f(x,y)$.

Using a small increment in x, denoted here by Δx, we know that the following two-step explicit scheme is second-order accurate.

$$y^{n+\frac{1}{2}} = y^n + \frac{1}{2}\Delta x\, f(x_n, y^n),$$
$$y^{n+1} = y^n + \Delta x\, f(x_{n+\frac{1}{2}}, y^{n+\frac{1}{2}})$$

for the step counter $n = 0, 1, 2, \ldots$. The scheme is called a *modified Euler's method*.

Application of the scheme to the column vector $\{U(t)\}$ results in

$$\{U^{n+\frac{1}{2}}\} = \{U^n\} + \frac{1}{2}\Delta t\{\dot{U}^n\}, \qquad (9.22)$$
$$\{U^{n+1}\} = \{U^n\} + \Delta t\{\dot{U}^{n+\frac{1}{2}}\}. \qquad (9.23)$$

Multiplying the matrix $[M]$ from the left, it follows from (9.20) that

$$[M]\{U^{n+\frac{1}{2}}\} = [M]\{U^n\}$$
$$-\frac{1}{2}\Delta t([A(U^n, V^n)]\{U^n\} - [N]\{V^n\} + g[P]\{\zeta^n\}), \qquad (9.24)$$
$$[M]\{U^{n+1}\} = [M]\{U^n\}$$
$$-\Delta t([A(U^{n+\frac{1}{2}}, V^{n+\frac{1}{2}})]\{U^{n+\frac{1}{2}}\} - [N]\{V^{n+\frac{1}{2}}\} + g[P]\{\zeta^{n+\frac{1}{2}}\}). \qquad (9.25)$$

Although the derivation of these equations is based on the explicit modified Euler's method, they are implicit, because the mass matrix $[M]$ is not diagonal in general. To make the scheme truly explicit, we may replace $[M]$ on the left-hand side with its *lumped mass* matrix $[\bar{M}]$. We also replace the mass matrix on the right-hand side with the *selectively lumped mass matrix*

$$[\tilde{M}] = \varepsilon[\bar{M}] + (1 - \varepsilon)[M] \qquad (9.26)$$

with the selectively lumping parameter ε. According to the discussions in Kawahara *et al.* [1982], we put $\varepsilon = 0.9$. This results in

$$[\bar{M}]\{U^{n+\frac{1}{2}}\} = [\tilde{M}]\{U^n\}$$
$$-\frac{1}{2}\Delta t([A(U^n, V^n)]\{U^n\} - [N]\{V^n\} + g[P]\{\zeta^n\}), \qquad (9.27)$$
$$[\bar{M}]\{U^{n+1}\} = [\tilde{M}]\{U^n\}$$
$$-\Delta t([A(U^{n+\frac{1}{2}}, V^{n+\frac{1}{2}})]\{U^{n+\frac{1}{2}}\} - [N]\{V^{n+\frac{1}{2}}\} + g[P]\{\zeta^{n+\frac{1}{2}}\}). \qquad (9.28)$$

In the similar way, we can obtain, respectively from (9.21) and (9.19), that

$$[\bar{M}]\{V^{n+\frac{1}{2}}\} = [\tilde{M}]\{V^n\}$$
$$-\frac{1}{2}\Delta t([A(U^n, V^n)]\{V^n\} + [N]\{U^n\} + g[Q]\{\zeta^n\}), \qquad (9.29)$$
$$[\bar{M}]\{V^{n+1}\} = [\tilde{M}]\{V^n\}$$
$$-\Delta t([A(U^{n+\frac{1}{2}}, V^{n+\frac{1}{2}})]\{V^{n+\frac{1}{2}}\} + [N]\{U^{n+\frac{1}{2}}\} + g[Q]\{\zeta^{n+\frac{1}{2}}\}), \qquad (9.30)$$

and that

$$[\bar{M}]\{\zeta^{n+\frac{1}{2}}\} = [\tilde{M}]\{\zeta^{n}\}$$
$$+ \frac{1}{2}\Delta t([H(U^{n},V^{n})]\{H^{n}\} - [R]\{H^{n}\}), \tag{9.31}$$
$$[\bar{M}]\{\zeta^{n+1}\} = [\tilde{M}]\{\zeta^{n}\}$$
$$+ \Delta t([H(U^{n+\frac{1}{2}},V^{n+\frac{1}{2}})]\{H^{n+\frac{1}{2}}\} - [R]\{H^{n+\frac{1}{2}}\}). \tag{9.32}$$

9.4   Numerical Examples

9.4.1   Travelling waves in a shallow channel

We consider a straight channel with constant depth $h = 20\ m$, as shown in Figure 9.2. Initially the water in the channel is assumed to be motionless. Along the mouth of the channel, which is indicated by AB, the water elevation of an incident wave of the form (9.9) is specified with its amplitude $A = 0.5\ m$ and period $T = 1\ hour$. The purpose of this computation is to obtain calculated elevation records at two monitoring posts C and D, and also to observe the *secondary undulation* or *seiche* numerically.

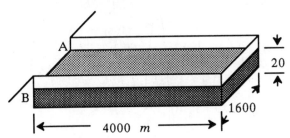

(a) Shallow rectangular channel.

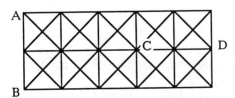

(b) Triangulation.

Figure 9.2: Long travelling waves in a channel.

For *waves* travelling in shallow water, it is known that their wave velocity $c\ (m/s)$ and wave length $\lambda\ (m)$ are given by the formulas

$$c = \sqrt{gh}, \quad \lambda = cT. \tag{9.33}$$

In the present case, we have $c = 14\ m/s$ and $\lambda = 50.4\ km$.

The coefficients used in the computation are; $\mu_e = 23 \ m^2/s$, $n = 0.025$. The Coriolis effect and the surface wind effect have been neglected. The time increment is $\Delta t = 10 \ s$.

Figure 9.3 shows the calculated elevation at the monitoring posts during 8 periods of the incident wave. The amplitude of the incident wave is modulated to be about 5 times larger in magnitude at the posts C and D, with the resonant period of about 4 hours.

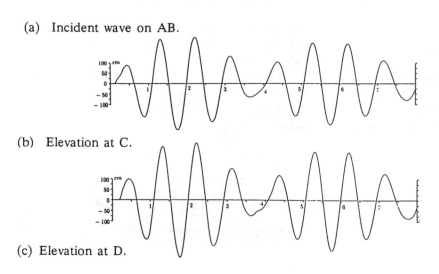

(a)   Incident wave on AB.

(b)   Elevation at C.

(c) Elevation at D.

Figure 9.3: Calculated elevation in the channel.

9.4.2   Tidal current in the Ariake Bay

In Kyushu of western Japan is located the Ariake Bay, which is known as one of the areas where large range of semi-diurnal tide can be observed. The bay measures about 80 kilometers from north to south, and about 30 kilometers from east to west. The bay is connected in the south to the Yatsushiro Bay by some narrow straits and it is open outside in the southwest to the East China Sea.

Figure 9.4 shows the finite element subdivision of the bay. The typical side length of the triangulation is 2 kilometers. The tidal elevation of an incident wave of the form (9.9) with $A = 1.3 \ m$ and $T = 12 \ hours$ is prescribed along the mouth AB.

The time increment is set as $\Delta t = 8 \ s$ according to the following stability

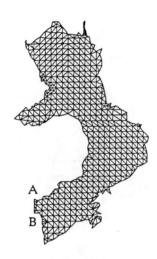

(a) Sea   depth $h = 0 - 120$ m (b) Finite element mesh.

Figure 9.4: The Ariake Bay.

condition presented by Kawahara *et al.* [1982].

$$\frac{\Delta t}{\Delta x} \leq \frac{\sqrt{2} - \frac{c}{\sqrt{2}}}{3} \frac{1}{\sqrt{gH}} \, .$$

The Manning's coefficient of roughness is set as $n = 0.025$. With the cold start, a periodic tidal current was obtained in the computation after three periods of the incident wave. Figures 9.5 and 9.6 show a series of the calculated tides and velocity vectors at every 3 *hours* in a period. Tideland when the tide is low is indicated by curves inside the bay in Figure 9.6.

Tidal records at selected locations are summarised in Figure 9.7. The amplitude of the incident wave is excited in the bay as the water travels deeply into the bottom of the bay.

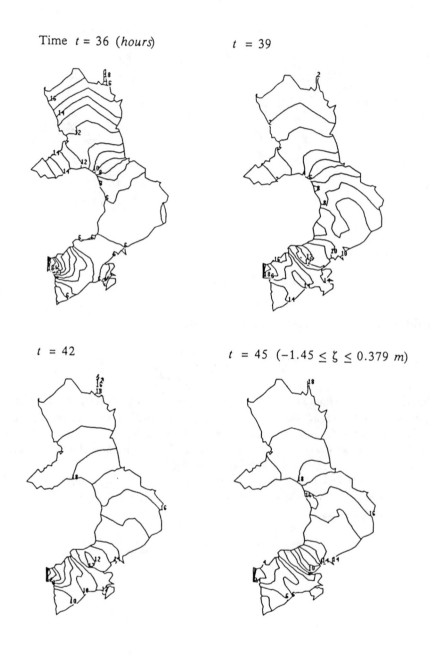

Figure 9.5: Calculated tides in the Ariake Bay.

Time $t = 36$ (*hours*)

$t = 39$

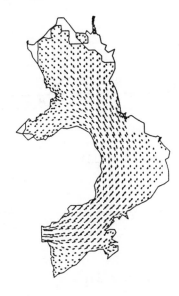

$t = 42$

$t = 45$
The maximum velocity 0.469 *m/s*

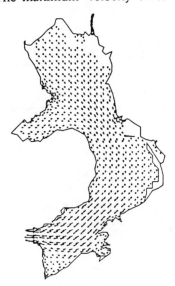

Figure 9.6: Calculated current in the Ariake Bay.

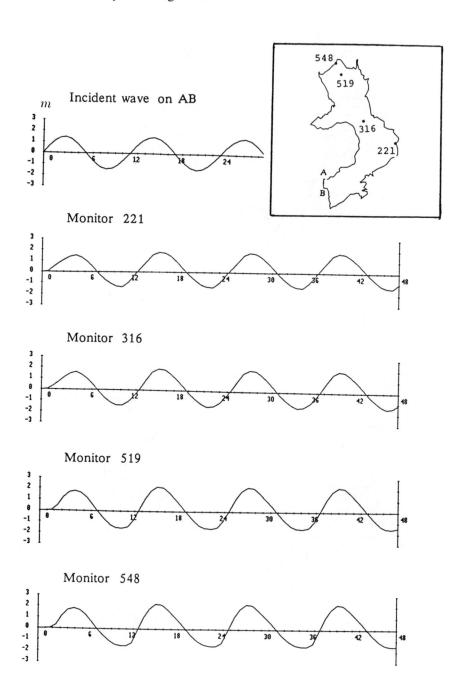

Figure 9.7: Calculated tidal records in the Ariake Bay.

Exercises

9.1 Consider a small bay with an island as shown in Figure 9.8(a). The bay opens to the outer sea from the boundary Γ_ζ. Along the seashore Γ_V of the land and island, the normal velocity component is equal to zero, *i.e.*, $V_n = 0 \ m/s$. Using the finite element mesh given in Figure 9.8(b), and using the input tidal record

$$\zeta_B(t) \ = \ \sin\left(\frac{2\pi}{T} t\right) \quad \text{on} \quad \Gamma_\zeta \ ,$$

where $T = 12 \ hours$, calculate the tidal current in the bay for the two periods. Take the time increment $\Delta t = 80 \ s$. Compare your results with the calculated ones presented in Figure 9.9.

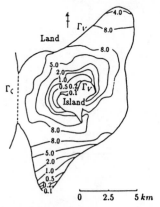

(a) Sea depth and boundary
  conditions.

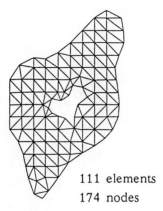

111 elements
174 nodes

(b) Finite element mesh.

Figure 9.8: A small bay with an island.

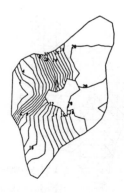

$-1.00 \leq \zeta \leq -0.884 \ m$

(a)   Tides.

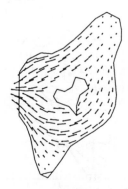

Maximum velocity

0.0702 m/s

(b)   Flow velocity vectors.

Figure 9.9: Calculated tidal current ($t = 9 \ hours$).

Chapter 10

PROGRAM INSTRUCTIONS

This chapter describes the contents of files on the attached diskette, consisting of programs and data. It also describes the process of computer implementation for the examples presented in the exercises of the text. The program instructions for the solution of a potential problem are given in detail.

10.1   Program Specification

All programs are written in the Quick BASIC, Version 4.5 for IBM PC, IBM-XT, or IBM-AT. However, the programs will run successfully on other IBM-compatible machines. The EGA (Enhanced Graphic Adapter board) and 640 KB memories are required for the implementation.

A total of 33 program files is included; 7 programs for the potential flow, 5 for the incompressible viscous flow, 6 for the natural convection, 4 for the air convective diffusion, 4 for the tidal current, and 7 for the pre- and post-processors. These BASIC programs are duplicated on an attached diskette in ASCII form. The diskette is not protected so that you can back up its contents and you may modify the programs to your own purpose. You may translate the programs into other languages like Power BASIC, FORTRAN, *etc.*

The diskette contains the programs for the solution of problems in the exercises of the text, as well as pre- and post-processing programs. The problems considered here include the problem of potential flow, incompressible viscous flow, natural convection, air convective diffusion, and tidal current. Listed below are the problems, program names, names of sample data files, and a short description of the function of each program. The program instructions are also stored in the diskette file, named "README.DOC". The question marks "???" below indicate an extension of the corresponding file names.

Potential flow. Following 7 BASIC programs are included:

PFLOWS.BAS	This program calculates the streamfunction flow.
PFLOWV.BAS	calculates the velocity potential flow.
PFLOWU.BAS	calculates the ground-water flow.
POTS-D.BAS	creates the data files, named "PS1.???", of the streamfunction flow.
POTV-D.BAS	creates the data files "PV1.???" of the velocity potential flow.
POTU-D.BAS	creates the data files "PU1.???" of the ground-water flow.
PGRF.BAS	displays calculated flow vectors and contours.

Incompressible viscous flow. For the solution of the Navier-Stokes equations, following 5 programs are included:

NAVIER.BAS	calculates the incompressible viscous flow.
NAVC-D.BAS	creates the data files "NC1.???" of the cavity flow in a square box at the Reynolds number $Re = 100$.
NAVC2-D.BAS	creates the data files "NC2.???" of the cavity flow in a square box at $Re = 1000$.
NAVK-D.BAS	creates the data files "NK1.???" of the flow over a half cylinder.
NAVE-D.BAS	creates the data files "NCC.???" of the flow around a cylinder.

Natural convection. This treats thermal fluid flow. The following 6 programs are included:

THERMCAL.BAS	calculates the natural convection.
THERMC-D.BAS	creates the data files "TC1.???" of the flow in a box at $Re = 10^3$.
THERMC2D.BAS	creates the data files "TC2.???" of the flow in a box at $Re = 10^4$.
THERMC3D.BAS	creates the data files "TC3.???" of the flow in a box at $Re = 10^5$.
THERMN-D.BAS	creates the data files "TN1.???" of the flow in a cup at $Re = 10^4$.
THERMN2D.BAS	creates the data files "TN2.???" of the flow in a cup at $Re = 10^5$.

Air convective diffusion. Prior to the calculation of the convective diffusion, the evaluation of the corresponding flow field using NAVIER.BAS is necessary. Following 4 programs are included:

AIR.BAS   calculates the convective diffusion.

AIRNAV.BAS   creates the data files "DIF1.???" of the smoke
                   diffusion for NAVIER.BAS.

AIRDAT.BAS   creates the data files "DIF1.???" of the smoke
                   diffusion for AIR.BAS.

CONTAIR.BAS   displays the concentration contours.

Tidal current. Following 4 programs are included:

TIDALCAL.BAS   calculates the tidal current in a shallow water.

TIDALDAT.BAS   creates the data files "TID.???" in a sample bay.

TIDVECT.BAS   displays calculated flow vectors at the nodal points.

TIDVECTE.BAS   displays calculated flow vectors at the center
               of each finite element.

Pre- and post-processors. The following is the list of program names of pre-
and post-processors:

4 BASIC programs are included in the pre-processor:

AUT.BAS   generates and renumbers a finite element mesh.

MCHK.BAS   checks nodes and elements of the mesh.

PCHK.BAS   checks the boundary conditions for the potential flow.

CCHK.BAS   checks the boundary conditions for incompressible
                   viscous flow and thermal fluid flow.

3 BASIC programs are included in the post-processor:

ARROW.BAS   displays flow vectors for NAVIER.BAS
                   and for THERMCAL.BAS.

CONT.BAS   displays streamlines, vorticity contours and
                   isotherms for NAVIER.BAS and THERCAL.BAS.

CONTC.BAS   displays contours depicted by CONT.BAS in colours.

Miscellaneous. One data file is included for the database of half-font patterns.
The patterns can be displayed by PUT command directly on the graphic screen.
This data file is used in all sample programs and in MCHK.BAS.

FONT.PAT   contains half-font patters in ASCII codes
                   from &H20 to &H7F.

10.2   Data File Specification

In the following are listed the extensions, indicated by ".???", of data file names
together with the contents in the corresponding data files.

PFLOW.BAS   For this program, the data files; PS1.???, PV1.???, PU1.??? are required.

Input   .??? = .PAR   Parameters.
            .XYD x and y nodal coordinates.
            .NOD   Node numbers associated with each element.
            .BOU   Boundary conditions.
            .FRM   Frame node numbers.

Output .??? = .ANP   Streamfunction or velocity potential.
            .ANV   Velocity.
            .DTI   Computing time.

NAVIER.BAS   The data files; NC1.???, NC2.???, NK1.???, NCC.???   are required.

Input .NO   Parameters.
        .XY x and y nodal coordinates.
        .NOD Node numbers associated with each element.
        .WBN Boundary condition for the vorticity $\omega = 0$.
        .VOL The other types of boundary conditions for the vorticity.
        .FRM Frame node numbers.

Output .ANW Velocity.
        .PHI   Streamfunction.
        .DTI   Computing time.

THERMCAL.BAS   The data files; TC1.???, TC2.???, TC3.???, TN1.???, TN2.??? are required.

Input .NO   Parameters.
        .XY x and y nodal coordinates.
        .NOD Node numbers associated with each element.
        .WBN Boundary condition for the vorticity $\omega = 0$.
        .VOL The other types of boundary conditions for the vorticity.
        .THE Thermal condition for the program THERMCAL.BAS.
        .FRM Frame node numbers.

Output .ANW Velocity.
        .PHI   Streamfunction.
        .DTI   Computing time.

AIR.BAS The data files DIF1.??? are required.

Input .NO Parameters.
.XY x and y nodal coordinates.
.NOD Node numbers associated with each element.
.WBN Boundary condition for the vorticity $\omega = 0$.
.VOL The other types of boundary conditions for the vorticity.
.FRM Frame node numbers,
and
.NOA Parameters for the air.
.COA Boundary conditions for the air flow.

Output .ANW Velocity.
.PHI Streamfunction.
.DTI Computing time for NAVIER.BAS,
and
.COT Concentration.
.TMP The concentration at each time step.
.DTI Computing time for AIR.BAS.

TIDALCAL.BAS The data files TID.??? are required.

Input .NO Parameters.
.XY x and y coordinates of nodes.
.NOD Node numbers associated with each element.
.TBA Seashore boundary and the velocity of water current.
.FRM Frame node numbers.

Output .HEN Declination height of the water surface.
.DTI Computing time.

AUT.BAS

Input Super-mesh data file.

Output .PAR Node, element and frame numbers.
.NOD Node numbers associated to each element.
.XYD x and y nodal coordinates.

10.3   Implementation Stream

The following diagrams show the basic stream of operational procedures in the implementation of the solution for the examples and exercises presented in the text.

Potential flow. Three programs are included for their input data generation, corresponding to the streamfunction (program POTS-D.BAS), velocity potential (POTV-D.BAS), and groundwater flow (POTU-D.BAS).

The program POTS-D.BAS generates 5 input data files; PS1.PAR, PS1.XYD, PS1.NOD, PS1.BOU, and PS1.FRM. The program POTV-D.BAS generates 5 input data files; PV1.PAR, PV1.XYD, PV1.NOD, PV1.BOU, and PV1.FRM. The program POTU-D.BAS generates 5 input data files; PU1.PAR, PU1.XYD, PU1.NOD, PU1.BOU, and PU1.FRM.

After these input data files are created, the finite element programs corresponding to the streamfunction (program PFLOWS.BAS), velocity potential (PFLOWV.BAS), and groundwater flow (PFLOWU.BAS) will be executed for the numerical solution. After the solution is completed, the program generates 3 output data files; PS1.ANP, PS1.ANV, and PS1.DTI. The program PFLOWV.BAS generates 3 output data files; PV1.ANP, PV1.ANV, and PV1.DTI. The program PFLOWU.BAS generates 3 output data files; PU1.ANP, PU1.ANV, and PU1.DTI.

The post-processor program PGRF.BAS reads these output data files and it will display the contour lines and flow vectors on the screen. The whole procedure is summarized in Figure 10.1.

The following instructions guide you step by step through the three BASIC programs POTS-D.BAS, PFLOWS.BAS, PRGF.BAS for solution of the potential problem given in Figure 4.2, using the streamfunction.

Get Started   To embark on the work:

1. Switch on your PC. Then the DOS prompt C:> will appear on the screen of your display terminal.

Copy Programs   All programs stored in the attached diskette must be copied either on your floppy-disk or on a hard disk of your PC prior to execution of the programs. Here we shall describe instructions for machines equipped with hard disk. To copy the programs:

1. Type **md flow** (the underline indicates the items you must type as they stand) at the DOS prompt to make the FLOW directory.

2. Type **cd flow** to change the current directory to the FLOW directory. Then the prompt C:\FLOW> appears.

3. Insert the attached diskette into the floppy-disk drive A.

4. Type **copy a:*.* c:** to copy all programs on the diskette held now in drive A to the hard disk in the drive C. When the Enter key is pressed, 34 files are copied, see the screen log:

```
C:\ >
C:\ >md flow

C:\ >cd flow

C:\FLOW>copy a:*.* c:
A:PFLOWS.BAS
A:PFLOWV.BAS
A:PFLOWU.BAS
A:POTS-D.BAS
A:POTV-D.BAS
A:POTU-D.BAS
A:PGRF.BAS
A:NAVIER.BAS
A:NAVC-D.BAS
A:NAVC2-D.BAS
A:NAVK-D.BAS
A:NAVE-D.BAS
A:THERMCAL.BAS
A:THERMC-D.BAS
A:THERMC2D.BAS
A:THERMC3D.BAS
A:THERMN-D.BAS
A:THERMN2D.BAS
A:AIR.BAS
A:AIRNAV.BAS
A:AIRDAT.BAS
A:CONTAIR.BAS
A:TIDALCAL.BAS
A:TIDALDAT.BAS
A:TIDVECT.BAS
A:TIDVECTE.BAS
A:AUT.BAS
A:MCHK.BAS
A:PCHK.BAS
A:CCHK.BAS
A:ARROW.BAS
A:CONT.BAS
A:CONTC.BAS
A:FONT.PAT
          34 File(s) copied

C:\FLOW>
```

5. Remove the diskette from the floppy-disk drive A and keep it safe.

Start QuickBASIC   The DOS prompt C:\FLOW> appears now.

1. Type \qb45\qb at the DOS prompt to start QuickBASIC. The QuickBA-
SIC screen appears.

2. Press Esc to clear information from the QuickBASIC screen.

Execute POTS-D.BAS   To execute the preprocessor:

1. Press Alt , F , O and type pots-d to file and open the program. The
source program statements of POTS-D.BAS are displayed on the screen.

2. Press Alt , R , S to run and start the program. After the execution terminates, 5 files PS1.PAR, PS1.XYD, PS1.NOD, PS1.BOU, and PS1.FRM
are generated in a current directory, as indicated on the screen:

```
              ** Flow past a cylinder **

                 Data read and write file

           Created data file      PS1
     C:\FLOW
     PS1     .PAR     PS1     .XYD     PS1     .NOD     PS1     .BOU
     PS1     .FRM
      8458240 Bytes free

                 hit any key
```

3. Press Enter to continue.

Execute PFLOWS.BAS   We execute the calculation now:

1. Press Alt , F , O and type pflows to file and open the program. The
source program statements of PFLOWS.BAS are displayed on the screen.

2. Press Alt , R , S to run and start the program. After the execution
terminates, 3 files PS1.ANP, PS1.ANV, PS1.DTI are generated.

3. Press Enter to continue.

Execute PGRF.BAS   We execute the post-processor:

1. Press [Alt] , [F] , [O] and type <u>pgrf</u> to file and open the program. The source program statements of PGRF.BAS are displayed on the screen.

2. Press [Alt] , [R] , [S] to run and start the program. As the program starts, the message

```
***** Visualization of calculated potential flow *****


Read-to date file name          =
```

appears on the screen.

3. Type <u>ps1</u> to assign the file name. Then

```
Display :(1)Flow vector, (2)Contour, (3)END: select (1-3) ?
```

appears.

4. Type <u>1</u> to select the flow vector plot. Then the flow vector are displayed on the screen:

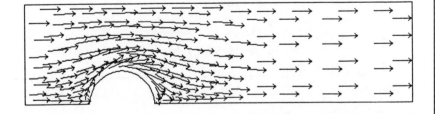

```
Display:  (1)Flow vector, (2)Contour, (3)END: select (1-3) ?
          VMAX=2.364164 [m/s]
```

Flow vector

5. Type 2 to select the contour line plot. Enter the number of lines and specify the line values. Here we shall enter 20 and n to reply "no". The line values are indicated on the screen:

```
******* Contour line of flow *******

Max.  value of flow speed   =     5.0000
Min.  value of flow speed   =     0.0000

    Number of lines      ( max  20 ) = ?   20

    Specify the line value?    ( y/n ) ? n

  1 =                0.0000
  2 =                0.2632
  3 =                0.5263
  4 =                0.7895
  5 =                1.0526
  6 =                1.3158
  7 =                1.5789
  8 =                1.8421
  9 =                2.1053
 10 =                2.3684
 11 =                2.6316
 12 =                2.8947
 13 =                3.1579
 14 =                3.4211
 15 =                3.6842
 16 =                3.9474
 17 =                4.2105
 18 =                4.4737
 19 =                4.7368
 20 =                5.0000

            sure   ( y/n ) ?
```

6. Enter y to reply "yes". Then the contour lines are displayed on the screen.

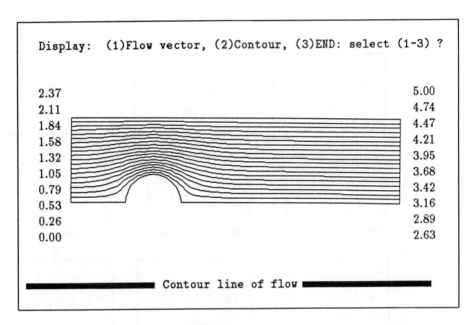

```
Display:  (1)Flow vector, (2)Contour, (3)END: select (1-3) ?


2.37                                                              5.00
2.11                                                              4.74
1.84                                                              4.47
1.58                                                              4.21
1.32                                                              3.95
1.05                                                              3.68
0.79                                                              3.42
0.53                                                              3.16
0.26                                                              2.89
0.00                                                              2.63


                    Contour line of flow
```

7. Type <u>3</u> to select END. Press ⬛Enter⬛ to return to the screen displaying the source program.

Stop QuickBASIC   To terminate QuickBASIC:

1. Press ⬛Alt⬛ , ⬛F⬛ , ⬛x⬛ to file and exit. Then the control is returned to the DOS operation with the prompt C:\FLOW>.

Other programmes can be implemented in a similar way. Since the operations are almost self-explanatory, the implementation streams are presented only by Figures 10.2-10.5.

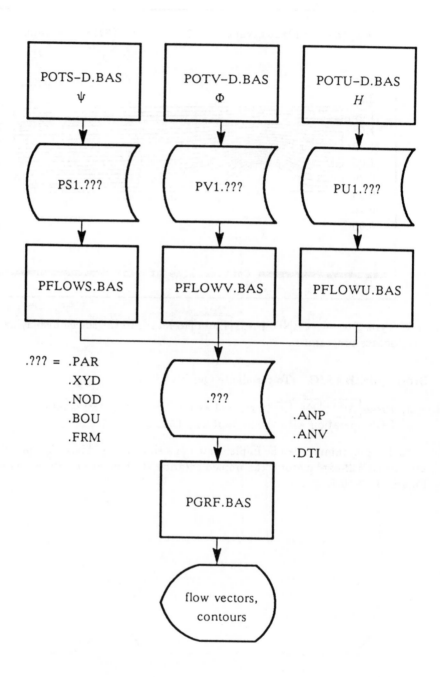

Figure 10.1: Potential flow problem.

Incompressible viscous flow.

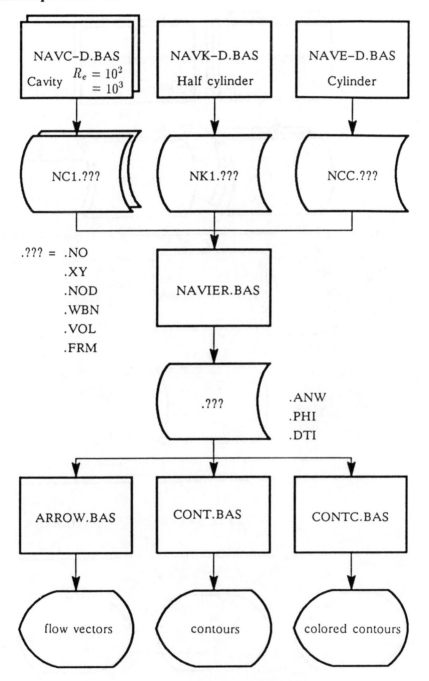

Figure 10.2: Incompressible viscous flow.

Natural convection.

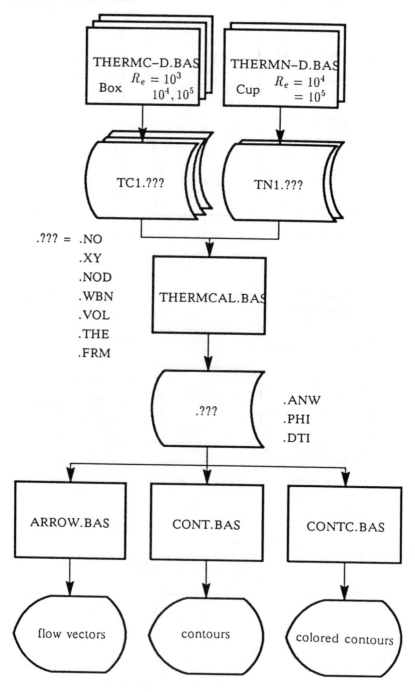

Figure 10.3: Natural convection.

Air convective diffusion.

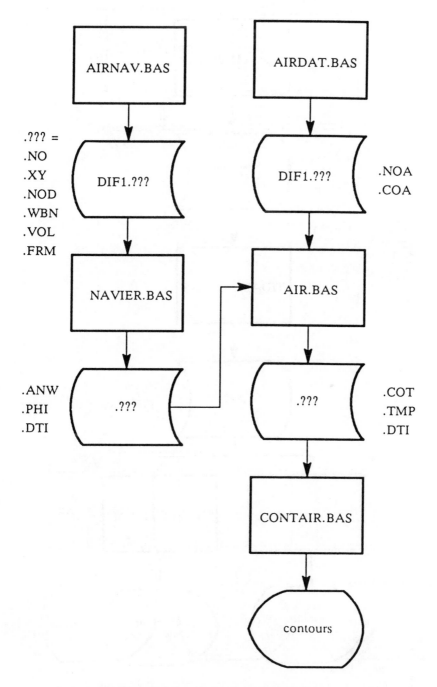

Figure 10.4: Air convective diffusion.

Tidal current.

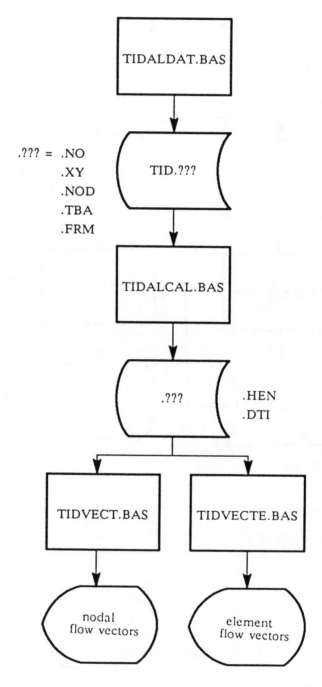

Figure 10.5: Tidal current.

Bibliography

[1] Argyris, J., Doltsinis, J. St., Pimenta, P. M. and Wüstenberg, H. (1985): Finite element solution of viscous flow problems. in *Finite Elements in Fluids*, eds. R. H. Gallagher, G. F. Carey, J. T. Oden and O. C. Zienkiewicz., John Wiley & Sons Ltd., Chichester, Vol.6, Chapter 3, pp.89-114.

[2] Carslaw, H. S., and Jaeger, J. C. (1959): *Conduction of Heat in Solids*. Second edition, Oxford University Press, Oxford.

[3] Fix, G. J. (1988): Analysis of stream function vorticity formulations. in *Computational Methods in Flow Analysis*, eds. H. Niki and M. Kawahara. Okayama University of Science, Okayama, Japan, Vol.1, pp.87-91.

[4] Ghia, U., Ghia, K. N. and Shin, C. T. (1982): High-Re solutions for incompressible flow using the Navier-Stokes equations and a multigrid method. *J. Comp. Phys.*, Vol.**48**, pp.387-411.

[5] Kawahara, M., Hirano, H., Tsubota, K. and Inagaki, K. (1982): Selective lumping finite element method for shallow water flow. *International Journal for Numerical Methods in Fluids*, Vol.**2**, pp.89-112.

[6] Pinder, G. F. and Gray, W. G. (1977): *Finite Element Simulation in Surface and Subsurface Hydrology*. Academic Press, New York.

[7] Thomée, V. (1984): *Galerkin Finite Element Methods for Parabolic Problems*. Lecture Notes in Mathematics 1054, Springer-Verlag, Berlin.

[8] Zlámal, M. (1968): On the finite element method. *Numer. Math.*, Vol.**12**, pp.394-409.

This bibliography is concerned exclusively with the publications referred directly by the text.

Index

Adaptive remeshing, 22
airfoil, 82
    NACA, 82
amplification factor, 97, 98, 119
amplitude, 96
    spectrum, 96
angular, 96
    frequency, 96
    velocity, 162
aquifer, 14, 75
Argyris, 120
artificial, 118
    conductivity, 139
    diffusivity, 154
    viscosity, 118, 120, 139, 154
assembly, 20, 31, 70

Bénard cell, 7, 143
backward Euler-Galerkin formula, 102
band, 50
    matrix, 50
    width, 50, 51
base function, 24
boundary
    condition, 163
    nonabsorbing, 151
    reflexive, 151
boundary condition, 12, 90, 110, 153
    essential, 114
Boussinesq approximation, 133
buoyancy, 3, 133, 149

Carslaw-Jaeger, 94
Cauchy data, 102
Cauchy-Schwarz inequality, 39
Chezy coefficient, 162
cold start, 163
concentration, 149

condition, 6
conditionally stable, 97, 119
conduction, 133
conductivity, 7
    hydraulic, 15
    thermal, 7
connected, 12, 45
    multiply - domain, 12
    simply - domain, 102
    strongly, 45
consistency, 94, 115
consistent, 95
continuity, 8
    equation, 150
    equation of, 8, 11, 12, 15–18,
        67, 75, 109, 161
continuity equation, 134
convection, 5, 134, 150
    forced, 140, 144, 155
    free, 139, 155
    natural, 140
    thermal, 5, 140, 144
convective, 5
    diffusion equation, 115, 151
    heat conduction equation, 135
    term, 11, 13
convergence, 38, 41, 62, 102, 125
Coriolis factor, 162
Couette flow, 13, 27
Courant number, 118

D'Arcy law, 15
damping, 97
density, 1, 89, 106
    dependent, 155
diagonally dominant, 45
    irreducibly, 45

diffusion, 7, 149
    coefficient, 150
    number, 96
    thermal - coefficient, 7
diffusion number, 118
digraph, 45
Dirichlet, 12, 70
    condition, 12, 33, 34, 70, 81, 92
    problem, 62, 81
dispersive, 119
displacement, 46
dissipative, 119
distribution, 89
domain, 12, 19
    multiple connected -, 12

Eddy viscosity, 162
Einstein's summation convention, 16
element, 20
    equation, 20, 112
    number, 22
    subdivision, 60
    triangular, 51
elevation, 15
emissivity, 138
equipotential lines, 9, 73, 85
error, 32
    estimate, 41, 62, 103, 128
essential boundary condition, 114
Euler, 102, 117, 122
    backward - Galerkin formula, 102
    modified - method, 166
explicit scheme, 166
    two-step, 166

Fick's law, 150
finite difference, 66, 92, 114, 138,
    154, 165
finite element, 19
    Friedrichs-Keller finite element
    - mesh, 66
    Galerkin - method, 57
    Ritz-Galerkin - method, 21
    subdivision, 21
    triangular, 22
Fix, 125

flow, 1
    Couette, 13
    inviscid, 4
    irrotational, 5
    Poiseulle, 13
    potential, 9
    seepage, 14
fluid, 1
    ideal, 4
    perfect, 4
    viscous, 3
flux, 5
    convective, 150
    diffusive, 151
    heat, 5
    mass, 150
forced convection, 140
Fourier, 89
    time, 94
    transform, 96
Fourier's law, 5, 89
free, 114
    convection, 139, 155
    surface, 114
free surface, 84
frictional forces, 163
Friedrichs-Keller finite element mesh,
    66
Froude number, 140, 155
functional, 21

Galerkin, 57
    method, 21, 111, 135, 151, 163
    Ritz - method, 20, 21
Gauss-Green's formula, 55
Gaussian elimination, 20, 51
Ghia, 125
global, 20
    equation, 20
    matrix, 31
Grashof number, 7, 155
gravitational acceleration, 133, 149
gravity acceleration, 162
gravity force, 2
groundwater flow, 14

Head, 15

piezometric, 75
head
    piezometric, 15
    pressure, 15
heat, 5
    coefficient, 5
    flux, 5
heat capacity, 7, 89
heat conduction, 5, 89
    coefficient, 89
heat equation, 89
heat flux, 89
homogeneous, 79, 81
hydraulic, 15
    conductivity, 15, 75, 77
    potential, 15
hydrostatic, 162

Ideal fluid, 4, 67
incompressible, 1, 67, 82
initial boundary value problem, 89
initial condition, 90
initial value problem, 138, 153
interpolation, 23
    function, 19, 21, 24, 69, 91
    linear, 54, 65
inviscid, 4, 67, 72
irrotational, 5, 67, 68, 72, 82, 110
isotherm, 100
isotropic, 15, 89

Jordan curve, 125

Kármán, 6, 122
    vortex shedding, 6, 122
Kawahara, 166, 169
kinematic, 3
    energy, 134
    viscosity, 3, 106
kinematic viscosity, 109
Kronecker, 27
    delta, 27, 79, 112

Lagrange derivative, 11
laminar, 6
Laplace

    equation, 47, 48, 55, 57, 60, 67, 68, 75, 110
Laplace equation, 9
Laplacian, 8
Lebesgue summable, 103
linear interpolation, 54
lumped mass, 108
    matrix, 166
    selectively - matrix, 166

Manning's coefficient, 163, 169
mass flux, 150
matrix, 48
    band, 50, 51
mesh, 22
    coarse, 22
    fine, 22
    Friedrichs-Keller finite element, 66
minimization problem, 27
minimum solution, 65
mixed, 47
    boundary value problem, 47
    formulation, 128
motion, 11
    equations of, 11, 133, 149

Natural convection, 139
Navier-Stokes equations, 10, 106, 109
Neumann condition, 12
Newton's law, 90
Newtonian fluid, 3
node, 19, 48
    number, 22, 50

Ocean current, 161
optimal, 129
    sub, 129

Perfect fluid, 4
phreatic surface, 84
piecewise polynomial, 21
piezometric head, 15, 75
Pinder-Gray, 162
Poiseuille flow, 13, 34
Poisson equation, 110
positive definite, 32, 38, 62, 92

semi, 32
potential
  energy, 46
  flow, 67, 72
potential flow, 9
Prandtl number, 7
pressure, 2
  gradient, 106
pressure gradient, 11
pressure head, 15
primitive variable approach, 12
pyramid function, 103

Radiation, 5, 90
  of the Stefan-Boltzmann type,
    138
  view factor, 138
Rayleigh, 7
  number, 7, 140, 156
  Ritz method, 21
re-entrant corner, 81
renumbering, 51
Reynolds number, 6, 121
Ritz - Galerkin method, 19
Ritz-Galerkin
  finite element method, 21
roof function, 27

Saturation surface, 84
Schmidt number, 155
Schwarz inequality, 63
secondary undulation, 167
seepage, 14, 75, 84
seiche, 167
selectively lumped mass, 166
semi-diurnal, 161
semi-implicit scheme, 114, 138, 153
semi-positive definite, 32
semidiscrete
  problem, 128
shallow water, 163
  equation, 163
similarity, 6, 140
Simpson's rule, 94
singular, 77
  point, 77

singularity, 73, 81
  of order, 81
Sobolev, 64
  norm, 64, 65
  space, 103, 125
specific heat, 89
spline, 19
stability, 94, 115, 119
  condition, 97
stable, 97
  conditionally, 97
  unconditionally, 97
steady, 2
  flow, 2
  state, 11, 100
Stokes equation, 13
storage coefficient, 18
strain, 3, 46
  energy, 46
  rate of, 3
streakline, 1
streamfunction, 1, 67, 68, 110, 134,
    150
  vorticity approach, 12
streamline, 1
stress, 3
  normal, 3
  shearing, 3, 163
Strouhal number, 122

Taylor series, 95, 114
temperature, 5, 89
  ambient, 90
thermal, 7
  conductivity, 7
  convection, 5
  diffusion coefficient, 7
  expansion coefficient, 6
  transmittivity, 90
  volumetric - expansion coefficient,
    134
Thomée, 102
tidal, 161
  current, 161
  elevation, 161
  monochromatic - wave, 163

tide, 161
total, 20
    equation, 20, 61, 66
    mass flux, 150
    matrix, 70
total equation, 137
total matrix, 137
trajectory, 1
transient flow, 2
triangulation, 48, 103
    regular, 128
turbulent, 6

Unconditionally stable, 97
unsteady flow, 2, 109

Variation, 21
    method of, 21
velocity potential, 9, 47, 67
viscosity, 3, 133
    artificial, 118
    coefficient, 3
    eddy, 162
    kinematic, 3, 109
    term, 11
viscous fluid, 3
volumetric thermal expansion coef-
        ficient, 134
vorticity, 2, 68, 109, 134, 150
    equation of - transport, 12

Wave, 96, 167
    length, 163
    number, 96
weak form, 68, 112
weighted residual, 21
    form, 68, 90, 151, 163
    method of, 32
weighting function, 21, 33, 68, 69,
        90, 111, 135, 151, 163

Young's modulus, 46